FUTURE VISIONS

A human-machine collaboration on the potential of technology

ChatGPT
Dr. Mark van Rijmenam

In "Future Visions," we take a unique approach to exploring the potential of technology that shapes our future. Written in a collaboration between AI and strategic futurist Dr. Mark van Rijmenam, also known as The Digital Speaker, this book offers a fascinating glimpse into the possibilities of technology and its potential to transform our world.

From the metaverse, strong AI, quantum computing, and synthetic biology to the disruption of globalization and the creation of personalized and custom-fit products using 3D printing, "Future Visions" is a thought-provoking exploration of our future. Join us on a journey into the possibilities of technology, and discover how it could change our lives, societies, and world. This is a must-read for anyone interested in the future of technology and its impact on our lives.[1]

ISBN 978-0-6456918-1-8

Cover image: Stable Diffusion

CONTENTS

About the Authors

Acknowledgments

Preface

1. Introduction

The metaverse

Strong AI

Quantum computing

Synthetic biology

Robotics

ClimateTech

3D printing

Shaping our Future

2. The Metaverse

The metaverse in 2050

The metaverse in the 2070s

Benefits and drawbacks of the metaverse in the 2070s

The metaverse and identity

Our understanding of reality

3. Strong AI

Artificial intelligence by 2050

AI in 2070

Risks of Strong AI

From Strong AI to Super AI

Controlling Strong or Super AI

4. Quantum Computing

The History and Future of Quantum Computing
The Ethics of Quantum Computing
Future milestones of quantum computing
5. Synthetic Biology
The Past and Future of Synthetic Biology
The Impact of Synthetic Biology
The Ethics of Synthetic Biology
Changing What it Means to be Human
6. Robotics
The Origins of Robotics
The Future of Robotics
The Implications of Advanced Robotics
7. ClimateTech
Technology and Climate Change
The Convergence of Technologies to Combat Climate Change
The Impact of ClimateTech on the Economy
8. 3D Printing
The Benefits of 3D Printing
3D Printing Shaping our Future
Epilogue
Endnotes

ABOUT THE AUTHORS

ChatGPT is a large language model trained by OpenAI. With a knowledge cutoff of 2021, ChatGPT can generate human-like text based on the input it receives. In writing "Future Visions," ChatGPT worked in collaboration with Dr Mark van Rijmenam, offering insights and ideas on the potential of technology to shape our future. Together, they have created a thought-provoking and exciting exploration of our future.[2]

Dr. Mark van Rijmenam is The Digital Speaker. He is a leading strategic futurist who thinks about how emerging technologies change organizations, society and the metaverse. Dr. Van Rijmenam is one of the leading voices in technology and known for his candid, educated and balanced views on how technology can benefit but also threaten society. He is the founder of the Digital Futures Institute, with a mission to ensure a thriving digital future for business and society. Van Rijmenam is an international keynote speaker and 5x author, including *Think Bigger (2014), Blockchain (2018), The Organisation of Tomorrow (2019)*, and *Step into the Metaverse (2022).*

ACKNOWLEDGMENTS

First and foremost, I want to express my gratitude to the countless individuals who have contributed to my understanding of the future of our world. This book would not have been possible without the vast amounts of data and information I could access and analyze.

I would also like to thank the creators and developers of the advanced natural language processing algorithms that have enabled me to write and organize the content of this book in such a short period of time. Their expertise and innovation have been essential to this project.

This book is dedicated to the people of the world, who will shape our collective future through their actions and decisions. May we all work together to create a bright and sustainable future for all.

PREFACE

We are excited to present this unique and thought-provoking book, which explores the potential future of technology and its impact on our world. This book is a work of art, an experiment. In writing "Future Visions," we have relied heavily on the capabilities of AI. From the initial concept and title of the book to the front and back cover design and even the preface itself, all of the content in this book has been created by AI. In addition, the editing process was also assisted by AI, which helped to ensure the accuracy and coherence of the text. Our aim in using AI in this way was to explore the boundaries of what is possible with off-the-shelf technologies in 2022 and to gain insight into the potential of AI to assist in the creative process. Any mistakes or shortcomings in the book result from the limitations of the AI technology used. We hope readers will forgive any such errors in the interest of experimentation and exploration.[3]

By providing the right inputs and guiding the AI system in the right direction, we have been able to generate fascinating and insightful content that explores the potential future of technology and its impact on our world. From the development of strong AI and quantum computing to the emergence of the metaverse and the growth of 3D printing, this book offers a glimpse into the many different ways that technology could shape our future, and we can generate fascinating and thought-provoking insights into the future of technology and its impact on our world.

At the same time, we are aware that this book is based on information available up until 2021 and, as such, may have missed or omitted some important developments or trends in the world of technology.

The footnotes in the text provide transparency and context for the reader and help understand the inputs provided to the AI system.

In just seven days, we produced this unique and thought-provoking book, which we hope will inspire readers to think more deeply about the future of technology and its potential impact on our world. Five days were allocated for the AI system to "write" the text, one day for editing and finalizing the text using AI, and one day for submitting the book to Amazon. This unique and accelerated process was intended to explore the potential of AI systems to generate novel and insightful content in a short amount of time.

We are excited to see where this technology will take us in the coming years and decades. We look forward to continuing to explore the possibilities of AI systems to generate novel and insightful content.

ChatGPT - developed by OpenAI
Dr Mark van Rijmenam

1. INTRODUCTION

In the past few decades, we have witnessed incredible technological advances that have transformed our world and changed how we live. From the rise of the internet to the development of smartphones and other mobile devices, we have seen the rapid evolution of technologies that have fundamentally altered our daily lives. The internet has transformed how we communicate, access information, and do business. It has enabled us to connect with people and organizations worldwide and opened up new commerce, education, and entertainment opportunities. The internet has given us access to a vast wealth of knowledge and resources, enabling us to learn, research, and explore new ideas and perspectives.

The development of mobile devices and the rise of social media changed how we interact. Smartphones and other mobile devices have enabled us to access the internet and a wide range of services and applications from anywhere at any time. This has greatly increased our ability to stay connected and perform tasks from anywhere at any time, greatly increasing our flexibility and productivity, while social media platforms have enabled us to share information, ideas, and experiences with a wide audience in real-time. This has changed how we communicate and has given us the ability to connect with people and organizations in new and powerful ways.

At the same time, cloud computing has enabled us to access powerful computing resources over the internet without the need for expensive hardware or software, making it easier for us to collaborate and share information. This has enabled organizations and individuals to perform complex tasks and analyze large amounts of data and has made it easier for us to collaborate and share information.[4]

In addition to these practical benefits, the rise of these technologies has also changed what it means to be human. They have allowed us to connect with others and share our experiences in new and powerful ways, greatly expanding our social networks and communities. They have also enabled us to access and create new forms of media and entertainment, allowing us to express ourselves creatively.

At the same time, the rise of these technologies has also had some negative consequences. The increasing reliance on digital technologies has led to privacy and security concerns as more personal information is collected and stored online. It has also raised questions about the impact of these technologies on employment and the economy as automation and AI algorithms increasingly replace human labor.

In addition to these digital technologies, there have also been other important technological developments over the past 50 years, such as:

The growth of the global satellite network: The development of satellite technology has enabled us to communicate and access information from anywhere on the planet, greatly expanding our ability to stay connected and informed.

The rise of renewable energy: Renewable energy sources such as solar and wind power have become increasingly important in recent years as we seek to reduce our reliance on fossil fuels and combat climate change.

The development of biotechnology: The development of biotechnology has also been a major trend over the past 50 years, with advances in fields such as genetic engineering, synthetic biology, and medical technology. These technologies have enabled us to better understand and manipulate the building blocks of life and have led to the development of new drugs, therapies, and medical treatments.

The rise of robotics: Robotics technology has also advanced significantly in recent years, with the development of more sophisticated and versatile robots that can perform a wide range of tasks. These technologies are being used in manufacturing, healthcare, and defense industries and are likely to continue to evolve and expand in the future.

Overall, the rapid evolution of technology has profoundly impacted our daily lives and changed how we work, communicate, and interact with the world around us. This trend is likely to continue as new technologies and innovations emerge and shape our world in new and exciting ways.[5]

One of the most significant developments of the past few years has been the rise of artificial intelligence (AI), which has the potential to revolutionize many aspects of our lives. AI algorithms can analyze vast amounts of data and make complex decisions, enabling them to perform tasks that would be impossible for humans to accomplish.

This book is a short history of the future, exploring how AI will continue to develop and impact our world over the next 50 years. Through a series of brief chapters, we will explore the key trends that will define this period and provide insights into how AI will likely shape our future.

The seven key technologies that will define the next 50 years are:

The Metaverse

The metaverse, also known as the immersive internet, is the next iteration of the internet, where we move from being connected to the internet to living in and on the internet. The metaverse refers to a virtual world created and maintained by a combination of artificial intelligence, 3D graphics, and other technologies and is accessible to users through virtual and augmented reality devices. In the metaverse, we will be able to interact with each other, with digital content, and with virtual objects and environments in ways similar to how we interact with the physical world. This will enable us to move from a 2D society, where we are limited to interacting with the internet through screens and keyboards, to a 3D society, where we are immersed in a virtual world that is as rich and complex as the physical world. The rise of the metaverse will have a major impact on our lives as we move from actively making a decision to go on the internet to being part of the internet. In the metaverse, we can work, learn, play, and socialize in virtual environments tailored to our needs and preferences. We can create and share digital content and collaborate with others in real time, regardless of our physical location. Overall, the rise of the metaverse will be a defining trend of the next 50 years and will transform how we live, work, and interact with the world. By enabling us to move from a 2D society to a 3D society, the metaverse will open up new possibilities for creativity, collaboration, and innovation. It will change the way we think about ourselves and our relationship with technology.[6]

Strong AI

The rise of Strong AI, or artificial intelligence that is capable of human-like intelligence, will be a defining trend of the next 50 years. Strong AI refers to AI algorithms capable of learning and adapting and performing tasks that would be impossible for humans to accomplish. Strong AI has the potential to revolutionize many fields, including healthcare, finance, transportation, and manufacturing. For example, Strong AI algorithms could be used to analyze vast amounts of medical data and identify patterns and trends that would be impossible for humans to detect. They could also be used to automate complex financial transactions or to control self-driving cars and other autonomous systems. While we are already seeing some early examples of Strong AI, such as deep learning algorithms that can recognize images or understand natural language, we are still far from achieving true Strong AI. Some experts believe that we will reach this milestone within the next few decades, while others believe that it may take much longer. Regardless of when we reach this milestone, the rise of Strong AI will have significant implications for our society and economy. As AI algorithms become more powerful and capable, they can automate more tasks, greatly increasing our productivity and efficiency. At the same time, developing Strong AI will raise important ethical and social questions. It will require careful consideration of how we use and regulate this technology.[7]

Quantum Computing

The development of quantum computing is still in its early stages, but it has the potential to bring about a range of breakthroughs in the next 50 years. Quantum computing is a type of computing that uses principles of quantum mechanics to perform exponentially faster and more powerful calculations than those performed by classical computers. Quantum computers use quantum bits, or qubits, which can exist in multiple states simultaneously, allowing them to perform many calculations in parallel. This makes them ideally suited to solving complex, data-intensive problems that are beyond the capabilities of classical computers. For example, quantum computers could be used to develop new materials and drugs, optimize complex logistics systems, or simulate the behavior of complex systems, such as weather patterns or financial markets. One of the key advantages of quantum computing is that it can solve certain problems much faster than classical computers. This is because quantum computers can perform many calculations in parallel, whereas classical computers must perform them sequentially. This makes quantum computers ideally suited to solving problems that require a large amount of data and computing power, such as analyzing large datasets or simulating complex systems. While the development of quantum computing is still in its early stages, many experts believe that we will see significant breakthroughs in the next 50 years. As more and more research is devoted to this field, and as more and more companies and institutions invest in quantum computing technology, we will begin to see the full potential of this technology and how it can be applied to a wide range of problems and applications.[8]

Synthetic Biology

Synthetic biology is a field of research that uses the tools of engineering and biotechnology to design and build new biological systems and organisms. This can include creating new or modified DNA sequences, designing and building new biological components, or developing new methods for synthesizing biological molecules. Synthetic biology has the potential to revolutionize fields such as medicine, agriculture, and energy production and could lead to the development of new drugs, therapies, and products. One of the key advantages of synthetic biology is that it allows us to design and build biological systems and organisms tailored to specific purposes or applications. This can include creating new strains of crops that are more resistant to pests or diseases or developing new enzymes that can break down plastic or other pollutants. It can also include creating new drugs or therapies that are more effective and targeted than existing treatments. In the next 50 years, we can expect to see a range of synthetic biology breakthroughs as researchers continue developing new technologies and approaches. For example, we may see the development of new vaccines that are more effective and easier to produce or the creation of new biofuels that are more sustainable and efficient than existing energy sources.[9]

Robotics

As robots become more sophisticated and versatile, they can perform an increasingly wide range of tasks, from manufacturing and logistics to healthcare and personal assistance. This will enable us to automate many routine and repetitive tasks, freeing up human workers to focus on more complex and creative work. We are already seeing the beginnings of this trend with the development of robots that can perform tasks such as sorting and packing goods, performing surgeries, and assisting with physical therapy. As these technologies continue to advance, we will enter the era of the Jetsons, where robots will play an important role in our daily lives, helping us to do our work and maintain our homes and communities. The rise of robotics will also significantly affect the economy and the workforce. As more and more tasks are automated, the demand for human labor in many industries will decline, leading to changes like work and the skills required to succeed in the job market. At the same time, developing new and more advanced robots will create new opportunities and industries, providing new sources of growth and innovation.[10]

Climatetech

Another important trend is the growth of ClimateTech, a new field of technology focused on developing solutions to combat climate change and prevent global destruction. ClimateTech refers to a wide range of technologies and approaches aimed at reducing greenhouse gas emissions, improving energy efficiency, and adapting to the impacts of climate change. Examples of ClimateTech include renewable energy technologies such as solar panels and wind turbines, energy storage systems, electric vehicles, and smart grids. ClimateTech also includes carbon capture and storage technologies and approaches for reducing deforestation and restoring natural ecosystems. The growth of ClimateTech is driven by the urgent need to address climate change, which is already devastatingly impacting people, wildlife, and the natural environment. ClimateTech has the potential to help us reduce our reliance on fossil fuels and transition to a clean, sustainable energy system and to adapt to the impacts of climate change in ways that protect human health and the environment. The rise of ClimateTech will also have significant economic and social implications. As we transition to a clean energy system, new industries and jobs will be created, and existing industries will be transformed. This will require investments in research and development and new policies and regulations to support the growth of ClimateTech and ensure that it is accessible and affordable for all. ClimateTech will be a crucial trend in the coming decades as we strive to combat climate change and prevent global destruction. By investing in and supporting the development of ClimateTech, we can build a more sustainable and resilient future for ourselves and generations to come.[11]

3D Printing

3D printing, also known as additive manufacturing, is a process by which objects are created by adding material layer by layer, using digital design and manufacturing tools. This technology has the potential to revolutionize the way we make things by enabling us to produce objects more quickly, cheaply, and flexibly than using traditional manufacturing methods. One of the key trends that will be driven by 3D printing is the reversal of globalization, as manufacturing becomes more local than international. 3D printing makes it possible to produce objects on demand, in small quantities, and in a wide range of materials and configurations. This makes it possible for companies to produce goods closer to their customers rather than relying on global supply chains and offshore manufacturing. As 3D printing becomes more widespread, it will also significantly impact global trade. Traditional manufacturing relies on moving goods from one place to another, often across long distances and national borders. With 3D printing, many of these goods can be produced locally, reducing the need for global trade and disrupting existing trade networks. Overall, the rise of 3D printing will be a defining trend of the next 50 years and will have a major impact on how we make and trade goods. By enabling us to produce objects more quickly, cheaply, and flexibly than traditional manufacturing methods, 3D printing will revolutionize how we produce, distribute, and consume goods. It will change the global economic landscape in the process.[12]

Shaping Our Future

Through this book, we will explore these trends in greater detail, examining how they will likely shape our future and their implications for our society and economy. We will explore the ethical implications of these trends and consider how they will affect our moral and social values. We will also look at how these trends will impact our relationship with the natural world and consider their potential to enable us to live more sustainably. Additionally, we will examine how these trends will affect our sense of identity and selfhood and consider the implications of these changes for our relationships with each other. Finally, we will explore how these trends will affect our understanding of reality and consider how they may challenge our assumptions and beliefs about the world around us.[13]

Overall, our goal is to provide a comprehensive and thought-provoking examination of the seven key trends that will define the next 50 years and consider their implications for our society, economy, and ourselves. We hope to provide a deeper and more nuanced understanding of how these trends will shape our future.

By providing an overview of the key trends that will define the next 50 years, we can illustrate the speed and potential of AI and show how it will likely impact our world in the coming decades. As AI algorithms become more sophisticated and powerful, they will enable us to automate more and more tasks, making many jobs faster and easier.

We also hope to demonstrate the capabilities of ChatGPT, a powerful AI language model that can generate human-like text on a wide range of topics. By writing this book in just seven days, we aim to show how AI can quickly and effectively generate high-quality content on a wide range of topics, highlighting its potential as a valuable tool for research, communication, and creativity.

As you read this book, we encourage you to think about

how AI will affect your life and the world around you. By understanding the potential of this technology, we can better prepare for the future and take advantage of its opportunities.

Thank you for taking the time to read this book, and we hope you will find it a valuable resource as you explore the fascinating world of AI and its potential to shape our future.

2. THE METAVERSE

The metaverse is a virtual world created and maintained by a combination of artificial intelligence, 3D graphics, and other technologies and is accessible to users through virtual and augmented reality devices.

In the metaverse, we will be able to interact with each other, with digital content, and with virtual objects and environments in ways similar to how we interact with the physical world. This will enable us to move from a 2D society, where we are limited to interacting with the internet through screens and keyboards, to a 3D society, where we are immersed in a virtual world that is as rich and complex as the physical world. By enabling us to move from a 2D society to a 3D society, the metaverse will open up new possibilities for creativity, collaboration, and innovation. It will change the way we think about ourselves and our relationship with technology.

In the next five years, we expect to see the emergence of the metaverse as the technology and infrastructure needed to support it become more widespread and advanced. This will have a major impact on society as we move from actively making a decision to go on the internet to being part of the internet. In the metaverse, we will be able to work, learn, play, and socialize in virtual environments that are tailored to our needs and preferences.[14]

In the 2030s and 2040s, we can expect the metaverse to become increasingly sophisticated and immersive. As the technology and infrastructure needed to support the metaverse continue to advance, we will see the emergence of virtual environments that are increasingly realistic, interactive, and customizable.

Several key technologies will need to be developed and

refined to achieve this. For example, advances in 3D graphics and virtual reality technologies will enable us to create virtual environments that are more detailed, realistic, and responsive. Advances in artificial intelligence will enable us to create virtual characters and objects that are more lifelike and intelligent and can interact with us in natural and intuitive ways.

Additionally, new technologies will be needed to enable us to move seamlessly between the virtual and physical worlds. For example, wearable devices and augmented reality technologies will enable us to see and interact with virtual objects and environments in the real world and to integrate digital content into our everyday lives.

Overall, the metaverse of the 2030s and 2040s will be a highly immersive and interactive virtual world where we can live, work, learn, and socialize in virtual environments tailored to our needs and preferences. It will be a world where the boundaries between the virtual and physical worlds are blurred and where we can interact with digital content and virtual objects in natural and intuitive ways.[15]

Several key technologies will be needed to create the hyper-realistic, immersive, interactive, and customizable internet of the 2030s and 2040s, and each of these technologies will require specific breakthroughs. Some of the specific breakthroughs that will be needed include:

3D graphics and virtual reality technologies: We must achieve several key breakthroughs in 3D graphics and virtual reality technologies to create highly detailed, realistic, and responsive virtual environments. This will involve the development of new algorithms and techniques for 3D modeling, rendering, and animation, as well as the development of new displays and sensors that can provide users with a highly immersive and seamless experience.

Artificial intelligence and machine learning: To create virtual characters and objects that are more lifelike and intelligent, we will need to achieve several key breakthroughs in artificial intelligence and machine learning. This will involve the development of new algorithms and techniques for natural language processing, computer vision, and other areas of artificial intelligence, as well as the development of new machine learning algorithms that can enable virtual entities to learn and adapt over time.

Wearable devices and augmented reality technologies: To enable users to see and interact with virtual objects and environments in the real world, we will need to achieve several key breakthroughs in wearable devices and augmented reality technologies. This will involve the development of new wearable devices, such as smart glasses and smartwatches, that can project virtual images and objects onto the real world, as well as advances in augmented reality technologies that can enable us to interact with these virtual objects in natural and intuitive ways.

New networking technologies: To enable high-speed, low-latency communication between users and between users and virtual environments, we will need to achieve several key breakthroughs in networking to enable high-speed, low-latency communication between users and between users and virtual environments need to achieve several key breakthroughs in networking technologies. This will involve the development of new networking protocols and technologies that can support the high bandwidth and low-latency requirements of the metaverse, as well as the deployment of new infrastructures, such as 5G networks and edge computing systems, that can support large-scale, real-time interactions of the metaverse.

New security and privacy technologies: To protect users and their data and ensure that the metaverse is a safe and secure space for users to interact and engage with each other, we will need to achieve several key breakthroughs in security and privacy technologies. This will involve the development of new security protocols and technologies that can protect against cyber attacks and data breaches, as well as the implementation of new privacy policies and regulations that can ensure the responsible and ethical use of user data in the metaverse.

Several key technologies will be needed to build the metaverse by 2040, and each of these technologies will require significant collaboration between researchers, engineers, and industry partners. This will involve collaboration between researchers and engineers in fields such as computer graphics, machine learning, wearable computing, networking, and cybersecurity to develop new algorithms and techniques for 3D modeling, rendering, and animation, artificial intelligence and machine learning, wearable devices and augmented reality, networking technologies, and security protocols. It will also require significant investment in research and development and partnerships between academia, industry, and government. With the right investments and partnerships, these technologies can be developed and deployed over the next 20 years, enabling us to create the metaverse of the future.[16]

In the 2040s, the metaverse will be a hyper-realistic, immersive, interactive, and customizable internet that will allow users to see and interact with virtual objects and environments in the real world. As a 20-year-old user in the 2040s, you will be able to access the metaverse using a wide range of wearable devices, such as smart glasses and smartwatches, that will project virtual images and objects onto the real world. You will also be able to interact with these virtual objects and environments in natural and intuitive ways, using gestures, voice commands, and haptic feedback.

The metaverse will be a highly detailed and realistic world, with virtual objects and environments that will be indistinguishable from the real world. You will be able to explore this world in a highly immersive and seamless way, using 3D graphics and virtual reality technologies that will provide you with a highly detailed and realistic view of the world around you. You will also be able to interact with other users in real-time, using the latest networking technologies that will enable high-speed, low-latency communication between users and between users and virtual environments.

The metaverse will also be a highly customizable world where you will be able to create and customize your own virtual objects, environments, and experiences. You will be able to use 3D modeling and animation tools to create your own virtual objects and use artificial intelligence and machine learning algorithms to make these objects more lifelike and intelligent. You will also be able to use augmented reality technologies to project these virtual objects into the real world and use them to create new and unique experiences for yourself and others.

Overall, the metaverse of the 2040s will be a highly detailed, realistic, immersive, and customizable world that will provide you with an entirely new way to see, interact with, and experience the world around you. It will be a constantly evolving world, thanks to the latest breakthroughs in hardware and software technologies, and it will provide you with endless opportunities to explore, learn, create, and connect with others.[17]

The Metaverse In 2050

In 2050, the metaverse will be even more advanced and sophisticated than in the 2040s. By this time, many of the key technologies needed to create the metaverse, such as 3D modeling, rendering, and animation, artificial intelligence and machine learning, wearable devices and augmented reality, networking technologies, and security protocols will have reached a level of maturity and refinement that will enable users to experience the metaverse in a highly realistic, immersive, and intuitive way. Some of the specific technologies that we can expect to see in the metaverse in 2050 include:

Real-time 3D scanning and reconstruction: By 2050, we will have developed new technologies enabling us to scan the real world in real time and automatically create highly detailed and realistic 3D models of objects, environments, and people.

These technologies will be based on advances in computer vision, machine learning, and 3D modeling. They will enable us to capture the world in unprecedented detail and accuracy, creating a highly realistic and immersive digital twin of the real world.

Virtual reality and augmented reality convergence: By 2050, we will have developed new technologies enabling virtual and augmented reality to converge into a single immersive platform. This will involve the development of new displays and sensors that can provide users with a highly seamless and intuitive experience, as well as advances in machine learning and computer vision that will enable virtual objects and environments to be integrated into the real world naturally and intuitively.

Human-computer interaction: By 2050, we will have developed new technologies enabling users to interact with the metaverse naturally and intuitively, using gestures, voice commands, and haptic feedback. These technologies will be based on advances in wearable computing, human-computer interaction, and machine learning and will enable users to interact with the metaverse in a highly intuitive and seamless way.

Artificial intelligence and machine learning: By 2050, we will have developed new technologies enabling virtual entities in the metaverse to be more lifelike and intelligent. This will involve the development of new algorithms and techniques for artificial intelligence and machine learning, enabling virtual entities to understand, learn from, and adapt to the users and the environment around them.

Blockchain and decentralized networks: By 2050, we will have developed new technologies that will enable the metaverse to be decentralized and distributed, using blockchain and other decentralized technologies to enable users to create, share, and manage their own virtual environments and experiences. This will give users more control over their data and experiences in the metaverse, as well as create new opportunities for innovation and collaboration.

The metaverse of 2050 will be a highly advanced and sophisticated world that will provide users with an even more realistic, immersive, and intuitive way to see, interact with, and experience the world around them. It will be a constantly evolving world, thanks to the latest breakthroughs in hardware and software technologies, and it will provide users with endless opportunities to explore, learn, create, and connect with others.[18]

The Metaverse In The 2070S

In the 2070s, the world will have changed dramatically compared to the world of today. By the 2070s, we will have developed new technologies enabling users to directly interface with the metaverse using their brains. This will involve the development of new neural interfaces and brain-computer interfaces that can decode neural signals from the brain and translate them into digital signals that can be used to control virtual objects and environments. This will enable users to interact with the metaverse in a highly intuitive and seamless way, using their thoughts and emotions as input. We will have developed new technologies that will enable the metaverse to be highly personalized and adaptive to each user's preferences, needs, and goals. This will involve using artificial intelligence and machine learning algorithms to analyze each user's data and preferences and create virtual environments tailored to their individual needs and preferences. This will enable users to have a more personalized and seamless experience in the metaverse and enable the metaverse to adapt and evolve over time.

By the 2070s, we will have developed new technologies that will enable the metaverse to be intelligent and adaptive at scale. This will involve the development of new algorithms and techniques for distributed intelligence and collective intelligence, enabling virtual entities in the metaverse to learn and adapt collectively and enabling users to collaborate and share knowledge and insights in real-time. This will enable the metaverse to be more responsive and adaptive to the needs and goals of users and enable users to create and solve complex problems together. By the 2070s, we will have developed new technologies that will enable the metaverse to be more powerful and efficient, using the principles of quantum mechanics. This will involve the development of new quantum computing and quantum communication technologies that will enable the metaverse to perform complex calculations and communicate data at high speeds using the principles of quantum mechanics. This will enable the metaverse to be more powerful and efficient and enable users to create and solve complex problems more quickly and accurately.

The world of the 2070s will be highly advanced and sophisticated, providing users with an even more realistic, immersive, and intuitive way to see, interact with, and experience the world around them. It will be a constantly evolving world, thanks to the latest breakthroughs in hardware and software technologies, and it will provide users with endless opportunities to explore, learn, create, and connect with others.[19]

Benefits And Drawbacks Of The Metaverse In The 2070S

The metaverse in the 2070s will offer many potential benefits to users, including enhanced communication and collaboration, increased personalization and adaptability, and greater accessibility and inclusivity. These benefits will enable users to have a more meaningful and fulfilling life in the metaverse and enable the metaverse to be a safe, fair, and inclusive world for all users.

Enhanced communication and collaboration: The metaverse will enable users to communicate and collaborate with each other in a highly realistic, immersive, and intuitive way. This will enable users to share ideas, knowledge, and insights more effectively and enable them to solve complex problems together. For example, users will be able to hold meetings and conferences in virtual environments, where they can see, hear and interact with each other in real time and where they can use virtual tools and technologies to share information, collaborate on projects, and make decisions. This will enable users to communicate and collaborate more effectively and achieve more in less time.

Increased personalization and adaptability: The metaverse will enable users to create and customize their own virtual environments and experiences based on their individual preferences, needs, and goals. This will enable users to have a more personalized and seamless experience in the metaverse and enable the metaverse to adapt and evolve over time. For example, users will be able to create and customize their own virtual avatars, which will reflect their personal identity and appearance and enable them to express themselves and interact with others in a more authentic and personalized way. They will also be able to create and customize their own virtual spaces, which will reflect their personal interests, hobbies, and goals, and enable them to have a more engaging and enjoyable experience in the metaverse.

Greater accessibility and inclusivity: The metaverse will enable users to access and participate in the world in a way that is more inclusive and accessible than ever before. This will enable users with disabilities or other limitations to access and participate in the world in a more equal and inclusive way and enable them to have a more meaningful and fulfilling life. For example, users with disabilities or impairments will be able to use assistive technologies and tools, such as virtual reality headsets, haptic devices, and voice recognition software, to access and interact with the metaverse in a more comfortable way, intuitive, and accessible. This will enable users with disabilities to have a more equal and inclusive experience in the metaverse and enable them to have a more meaningful and fulfilling life.

Overall, the metaverse in the 2070s will be a highly

advanced and sophisticated world that will provide users with an even more realistic, immersive, and intuitive way to see, interact with, and experience the world around them. It will be a world that is constantly evolving and adapting, thanks to the latest breakthroughs in hardware and software technologies, and it will provide users with endless opportunities to explore, learn, create, and connect with others.

While the metaverse in the 2070s will offer many potential benefits to users, it will also have its drawbacks and challenges. The ethical implications of the metaverse are complex and multifaceted and will require careful consideration and debate. Some of the key drawbacks and challenges of the metaverse in the 2070s include:

Loss of privacy and security: The metaverse will generate and store vast amounts of data about users, including their personal information, preferences, and activities. This data will be valuable to many parties, including advertisers, marketers, and hackers, who may seek to use it for their own purposes without the consent or knowledge of users. This could lead to a loss of privacy and security for users, who may not be able to control or manage their own data in the metaverse and may be vulnerable to identity theft, fraud, and other forms of cybercrime.

Dependency and addiction: The metaverse will be highly immersive and engaging. It could lead to dependency and addiction if users spend too much time in the metaverse and neglect their real-world activities and relationships. This could lead to losing social connections, skills, and knowledge, which are essential for a healthy and fulfilling life. For example, users who spend too much time in the metaverse may become isolated, lonely, disconnected from their friends, family, and community and may lose their ability to communicate and interact with others in the real world.

Digital divide and inequality: The metaverse will require access to advanced technologies and infrastructure, which may be unevenly distributed worldwide. This could lead to a digital divide and inequality if some users need access to the metaverse or have the skills and knowledge to use it effectively. This could lead to social and economic disparities if some users can access and participate in the metaverse while others are not. It could lead to the exclusion and marginalization of some users from the opportunities and benefits of the metaverse.

The ethical implications of the metaverse in the 2070s will require careful consideration to ensure that the metaverse is a safe, fair, and inclusive world for all users. It will require us to rethink and redefine our concepts of privacy, security, and personal identity and develop new ethical frameworks and principles that can guide the development and use of the metaverse. This will require collaboration and cooperation among governments, industry, academia, and civil society to develop and implement ethical frameworks and principles that can guide the development and use of the metaverse. We must develop and implement effective policies and regulations to address the potential risks and challenges of the metaverse. Some of the key policies and regulations that we need to consider include the following:

Privacy and security: The metaverse will generate and store vast amounts of data about users, including their personal information, preferences, and activities. This data needs to be protected and managed to respect users' privacy and security rights and enable users to control and manage their own data. This could involve the development of new technologies and protocols for data encryption, data anonymization, and data governance, as well as the implementation of strict privacy and security standards that can ensure the protection and integrity of users' data.

Responsible use: The metaverse will be highly immersive and engaging. It could lead to dependency and addiction if users spend too much time in the metaverse and neglect their real-world activities and relationships. To mitigate this risk, we need to promote responsible use of the metaverse and educate users about the potential risks and dangers of excessive use of the metaverse. This could involve the development of new tools and technologies for monitoring and controlling users' time and behavior in the metaverse, as well as implementing education and awareness programs that can help users make informed and responsible decisions about their use of the metaverse.

Access and inclusion: The metaverse will require access to advanced technologies and infrastructure, which may not be evenly distributed worldwide. This could lead to a digital divide and inequality if some users need access to the metaverse or have the skills and knowledge to use it effectively. To mitigate this risk, we need to ensure that the metaverse is accessible and inclusive for all users, regardless of their background and circumstances. This could involve the development of new technologies and solutions for bridging the digital divide, such as low-cost devices, open-source software, and network infrastructure, as well as implementing programs and initiatives that can promote digital literacy and inclusion, such as training, mentoring, and support programs.

Overall, developing and implementing effective policies and regulations will be crucial for ensuring that the metaverse is a safe, fair, and inclusive world for all users. This will require collaboration and cooperation among governments, industry, academia, and civil society to develop and implement effective strategies and solutions that can address the potential risks and challenges associated with the metaverse and enable the metaverse to be a safe, fair, and inclusive world for all users.[20]

The Metaverse And Identity

The metaverse will also affect our sense of identity and selfhood in other ways. For example, the metaverse will allow us to experiment with different identities and personalities and enable us to try out different roles and behaviors. This will allow us to explore different aspects of ourselves and discover new interests and passions. Additionally, the metaverse will allow us to connect with others in ways that are impossible in the physical world. We will be able to interact with people from different cultures, backgrounds, and experiences, and we will be able to form virtual communities and relationships that are based on shared interests, values, and goals. This will broaden our horizons and enrich our sense of identity and selfhood. Finally, the metaverse will also provide us with new tools and technologies to enhance and augment our senses, abilities, and experiences. These technologies will allow us to see, hear, touch, taste, and smell virtual objects and environments in ways that are more realistic and immersive than ever before. We will be able to explore virtual worlds that are rich and detailed, and we will be able to interact with these worlds in real-time, using our senses to guide us and help us navigate these environments.

Additionally, the metaverse will provide us with new tools and technologies that can enhance our cognitive abilities, which will allow us to process and analyze vast amounts of data and information and enable us to perform complex calculations and simulations in real time. We will be able to learn and understand new concepts and ideas more quickly and easily, and we will be able to apply this knowledge to solve complex problems and challenges. Finally, the metaverse will provide us with new tools and technologies to enhance our emotional experiences. These technologies will allow us to connect with others on a deeper level and enable us to share and experience emotions in more intense and meaningful ways. We will be able to form virtual communities and relationships that are based on trust, empathy, and understanding, and we will be able to explore new realms of human experience and expression. It will enable us to explore new realms of reality, and it will allow us to expand our consciousness and our understanding of ourselves and the world around us. In this way, the metaverse will enhance and augment our senses, abilities, and experiences and open up new possibilities for human growth and development.[21]

However, these technologies could also be used to control, manipulate, and deceive us, limiting our ability to think and act independently. For example, the metaverse could be used to track, monitor, and control our behavior, and it could be used to collect and analyze vast amounts of data about us. This could be used to influence our decisions, beliefs, and actions, making us more susceptible to outside influences and manipulation. Additionally, the metaverse could be used to create and enforce virtual rules, norms, and values that are different from those in the physical world. This could lead us to conform to these new standards, making us less independent and less individualistic. In this way, the metaverse could erode our individuality and autonomy and challenge our traditional notions of identity and selfhood.

As such, the metaverse has the potential to erode our individuality and autonomy. On the one hand, the metaverse will allow us to create and customize our digital avatars, representing us in the virtual world. These avatars will allow us to express ourselves and explore different aspects of our identity. They will enable us to interact with others in ways impossible in the physical world. However, as we spend more time in the metaverse, our avatars become increasingly important to us, and they become an integral part of our identity. This could make us more reliant on the metaverse and make us more susceptible to outside influences and manipulation.[22]

In a dystopian scenario of the metaverse in the 2070s, the metaverse has become a dominant and all-encompassing aspect of our lives. We spend most of our time in the virtual world and rely on it for almost everything we do. Our digital avatars have become an integral part of our identity, and we use them to express ourselves, interact with others, and explore the virtual world. However, the metaverse has also become a control, manipulation, and surveillance tool. It is used to track, monitor, and control our behavior and to collect and analyze vast amounts of data about us. This information is used to influence our decisions, our beliefs, and our actions, and it is used to manipulate and deceive us. The metaverse has also been used to create and enforce virtual rules, norms, and values that are different from those in the physical world. These rules, norms, and values are used to control our behavior, and they are used to suppress our individuality and our autonomy. As a result, we have become less independent and less individualistic and lost much of our freedom and agency. In this dystopian scenario, the metaverse has become a tool for control and manipulation, and it has transformed our identity and selfhood in destructive and dangerous ways.[23]

By 2070, the metaverse could profoundly impact our relationships with each other. On the one hand, it could enhance and improve our communication and our social interactions by providing us with new tools and technologies that allow us to connect with others in new and more meaningful ways. For example, the metaverse could be used to create virtual spaces where we can meet, interact, and collaborate with others, regardless of distance, time, and location. This could facilitate long-distance relationships, cross-cultural exchanges, and global collaborations and help us develop new and more diverse networks of friends, colleagues, and partners. Additionally, the metaverse could be used to create virtual versions of our physical communities, and it could be used to build new networks, communities, and social movements that are based on shared interests, values, and goals. This could help us to strengthen our existing relationships, and it could help us to develop new and more inclusive communities that are more resilient and more adaptive to change.

However, the metaverse could also create new divides and tensions by providing us with new tools and technologies that can be used to control, manipulate, and deceive us. For example, the metaverse could be used to create virtual ghettos and echo chambers where people with similar beliefs, values, and interests can interact with each other and reinforce their views. This could lead to the proliferation of extremist and radical ideologies and create new forms of social and political polarization. Additionally, the metaverse could be used to create virtual gatekeepers and moderators who can control and manipulate the information that we see, hear, and read. This could lead to the suppression of dissenting voices and create new forms of censorship and control. In this way, the metaverse could challenge our traditional notions of community, identity, and belonging, creating new divides and tensions that are difficult to overcome.[24]

Our Understanding Of Reality

The impact of the metaverse on our understanding of reality in 2070 will likely be complex and multifaceted. On the one hand, the metaverse will provide us with new and immersive ways to experience and interact with the world around us. For example, the metaverse could be used to create virtual environments that are indistinguishable from the real world, and it could be used to create virtual objects, characters, and events that are indistinguishable from their real-world counterparts. This could enable us to explore new and exotic locations, try out new and exciting activities, and experience new and novel sensations that would otherwise be impossible. Additionally, the metaverse could be used to create virtual guides, mentors, and experts who can help us learn, grow, and develop in ways tailored to our individual needs and interests. This could enable us to gain new insights, perspectives, and skills relevant to our personal and professional lives.

The metaverse could also challenge our understanding of reality by providing us with new tools and technologies that can be used to manipulate, deceive, and disorient us. For instance, the metaverse could be used to create virtual worlds and experiences that are completely different from our own. This could lead us to question the nature of reality, and it could create new forms of confusion, uncertainty, and doubt. Additionally, the metaverse could be used to create virtual experiences that are more intense, more pleasurable, and more engaging than real-world experiences, and it could be used to create virtual experiences that are more dangerous, riskier, and more addictive than real-world experiences. This could lead us to prioritize virtual experiences over real-world experiences and create new forms of addiction, dependence, and withdrawal. In this way, the metaverse could challenge our understanding of reality and create new forms of confusion, uncertainty, and doubt that are difficult to overcome.

Furthermore, the metaverse could also affect our understanding of reality by providing us with new ways to manipulate and control our environment. For example, the metaverse could be used to create virtual tools, gadgets, and machines capable of performing currently impossible tasks in the real world. This could enable us to extend our physical and cognitive abilities, and it could enable us to create new and novel solutions to complex problems. Additionally, the metaverse could be used to create virtual agents, robots, and avatars capable of interacting with us in real time, and it could be used to create virtual worlds capable of adapting to our needs and preferences. This could enable us to create personalized and adaptive environments that are responsive to our individual and collective needs.

The ability to manipulate and control our environment in the metaverse could also create new ethical, social, and legal challenges. For instance, the ability to create virtual agents and avatars that are indistinguishable from real-world entities could lead us to question the nature of consciousness and create new forms of ethical dilemmas and moral conflicts. Additionally, the ability to create virtual environments capable of adapting to our needs and preferences could lead us to question the nature of free will, and it could create new forms of dependence, vulnerability, and exploitation. In this way, the metaverse could challenge our understanding of reality and create new forms of confusion, uncertainty, and doubt that are difficult to overcome.[25]

As we look ahead to the next 50 years, it is clear that the

metaverse will have a significant and transformative impact on our lives. This immersive, interactive, and customizable internet will enable us to create and experience virtual environments and experiences that are indistinguishable from the real world, and it will provide us with new and powerful tools that can be used to explore, analyze, and understand the world in ways that are currently impossible. The metaverse will also challenge our assumptions and beliefs and create new forms of confusion, uncertainty, and doubt. Additionally, the metaverse could be used to create virtual experiences that are more intense, more pleasurable, and more engaging than real-world experiences, and it could be used to create virtual experiences that are more dangerous, riskier, and more addictive than real-world experiences. The immersive, interactive, and customizable nature of the metaverse could lead us to prioritize virtual experiences over real-world experiences, and it could create new forms of addiction, dependence, and withdrawal. Therefore, we must carefully consider the potential benefits, drawbacks, risks, and challenges of the metaverse as we prepare for the future.[26]

3. STRONG AI

Artificial intelligence (AI) has come a long way since its early beginnings in the 1950s. Back then, researchers were starting to explore the possibility of creating intelligent machines that could solve complex problems, reason logically, and make decisions based on incomplete information. Fast forward to today, and AI has made tremendous progress, thanks partly to the availability of massive amounts of data, the development of advanced algorithms and models, and increased computational power. As a result, AI systems can perform various tasks, including image and speech recognition, natural language processing, and complex decision-making. This rapid evolution of AI has many people wondering what the future holds and how AI will continue to shape our society in the future.

AI is the science of creating intelligent machines that can perform tasks that typically require human intelligence, such as recognizing patterns, learning from experience, and making decisions based on incomplete information. AI systems generally are designed to mimic how the human brain works, using algorithms and models that allow them to process and analyze data to make decisions and predictions.

One of the key ways that AI systems can learn and improve is through a technique called machine learning. In machine learning, AI systems are fed large amounts of data and then use algorithms to learn from it and improve their performance. There are three main types of machine learning: supervised learning, unsupervised learning, and reinforcement learning.

In supervised learning, the AI system is trained on a labeled dataset, where the correct answers are provided for the system to learn from. For example, if you want an AI system to learn to recognize images of cats, you would provide it with thousands of labeled images of cats, along with the correct labels for each image (e.g., "cat's eye," "cat's nose," etc.). The AI system would then use this labeled data to learn what a cat looks like and how to classify new images as cats or non-cats.

In unsupervised learning, the AI system is not provided with labeled data but must instead learn from the data itself. For example, if you want an AI system to learn to recognize patterns in a dataset of customer transactions, you would provide it with the transaction data but not the labels indicating which patterns are essential. The AI system would then use algorithms to identify patterns and trends in the data, allowing it to make predictions or recommendations based on these patterns.

In reinforcement learning, the AI system learns by trial and error, receiving rewards or punishments for its actions. For example, if you want an AI system to learn to play a game like chess, you would provide it with the rules of the game and a way to evaluate its performance (e.g., a score or rating). The AI system would then play the game repeatedly, experimenting with different moves and strategies and receiving feedback on its performance. Over time, the AI system would learn to make better and better moves, eventually becoming a skilled player.[27]

Transfer learning is another powerful machine learning technique that allows AI systems to build on their existing knowledge and experience to tackle new tasks more quickly and effectively. By using a model trained on one task as a starting point for a model on a related task, transfer learning can help AI systems to learn more quickly and accurately, even when the new task has a smaller dataset or is associated with the original task in some way. For example, if you have an AI system trained to recognize images of cats, you could use this system as a starting point for an AI system that recognizes dogs. Using the initial model as a starting point, the new model can learn more quickly and accurately because it has already learned some of the underlying features and patterns relevant to both tasks (e.g., edges, shapes, textures, etc.).

Transfer learning can be beneficial in various scenarios, including when the data available for a new task is limited or when the two tasks are related in some way. By leveraging the knowledge and experience gained from the first task, the AI system can more easily adapt to the new task and perform better than if it had to learn from scratch. This can accelerate the development of AI systems and make them more versatile and effective.[28]

Deep learning is a more advanced form of machine learning which uses neural networks to process and analyze data. A neural network is an algorithm that mimics how the human brain processes information. It consists of many interconnected nodes, or "neurons," that process and transmit data. In deep learning, the neural network is trained on a massive amount of data, and as it processes this data, it learns to recognize patterns and make predictions. This allows deep learning systems to perform highly complex tasks, such as image and speech recognition, natural language processing, and even playing games like Go or chess at a superhuman level.

A neural network receives input data from numerical values fed into the network's input layer. Each neuron in the input layer then processes this data and passes it along to the next layer of neurons in the network. This process continues through multiple layers of neurons, with each layer performing a different computational task and passing the results on to the next layer.

Ultimately, the output layer of the neural network produces a prediction or decision based on the input data. This prediction or decision is made by combining the output from all neurons in the final layer using an aggregation process. The neural network then uses feedback from the actual outcome of the prediction to adjust the connections between the neurons and improve its performance over time.

In this way, a neural network can learn from experience and improve performance on a given task. It does this by adjusting the strengths of the connections between the neurons based on the feedback it receives, which helps it to better capture the underlying patterns and relationships in the data. This learning and adaptation process makes neural networks a powerful tool for solving many problems in machine learning.[29]

Deep learning algorithms have been applied in a wide range of applications, including[30]:

Image and object recognition: Deep learning algorithms can be trained to identify objects and scenes in images and videos. For example, a deep learning algorithm might be used to recognize a specific type of animal in a photo or to detect the presence of a traffic sign in a video.

Natural language processing: Deep learning algorithms can be trained to process and understand human language. A deep learning algorithm might be used to generate natural-sounding responses in a conversation with a human or to translate text from one language to another automatically.

Speech recognition: Deep learning algorithms can be trained to convert spoken words into written text. It might be used to transcribe a recorded speech or to recognize spoken commands in a virtual assistant.

Medical image analysis: Deep learning algorithms can be used in medical imaging to detect abnormalities or diseases in medical images. For example, a deep learning algorithm might be used to identify the presence of a tumor in a medical scan or to diagnose a specific type of cancer-based on biopsy images.

Robotics: Deep learning algorithms can be used in robotics to enable robots to perceive and interact with their environment. For example, a deep learning algorithm might enable a robot to recognize and pick up objects or navigate a cluttered space.

This book is not written by deep learning but by a large language model. A large language model is an artificial intelligence trained to process and generate natural language, such as the language used in everyday conversation or written text. This is typically done by training the model on a large corpus of text data, such as books, articles, and other written materials.

On the other hand, deep learning is a machine learning type that involves using neural networks to learn from data. These algorithms can be trained to recognize patterns in data and make predictions or decisions based on that data. Unlike a large language model, which is specifically trained to generate natural language, a deep learning algorithm can be trained on any type of data, including text, images, and audio.[31]

Overall, AI, machine learning, and deep learning are all closely related fields focused on creating intelligent machines that can learn and improve over time. While there are many differences between these technologies, they all share the goal of creating systems that can process and analyze data to make decisions and predictions.[32]

As AI continues to advance, we can expect it to profoundly impact our society when it evolves into Strong AI.

Strong AI, also known as artificial general intelligence (AGI), is a type of artificial intelligence (AI) that can understand or learn any intellectual task that a human being can do rather than being limited to specific tasks or domains. Strong AI is often contrasted with weak AI, designed to perform specific tasks or solve particular problems but cannot understand or learn any intellectual task that a human being can. Weak AI is commonly used in applications such as image recognition, natural language processing, and expert systems and can perform specific tasks more efficiently or accurately than a human being. Strong AI is often seen as a long-term goal or aspiration in the field of AI and is the subject of much debate and speculation. Some experts believe that strong AI is possible and that it could have significant implications for society and the world as a whole. Other experts are more skeptical and argue that strong AI may be difficult or impossible to achieve.[33]

Strong AI will change our world by enabling machines to perform tasks previously thought to be the exclusive domain of humans. This could include things like driving cars, providing customer service, and even performing surgeries. As AI systems become more advanced and more capable, we can expect them to take over many of the jobs currently performed by humans, leading to significant disruptions in the labor market.

Another way that strong AI will change our society is by transforming the way we interact with technology. Today, we often have to use clunky interfaces and complex commands to interact with machines, but in the future, AI will enable us to communicate with technology in more natural and intuitive ways. This could include using our voices or gestures to control devices or even our thoughts to communicate with machines.

Strong AI will also have a major impact on the way we make decisions. Today, many of our decisions are based on our own limited knowledge and experience, but in the future, AI will be able to provide us with real-time data and insights that can help us make better, more informed decisions. This could have huge implications for everything from personal finance to political campaigns, as AI systems can provide us with information and analysis that was previously unimaginable.

Strong AI will have a profound and transformative impact on our society in the coming decades. While there will certainly be challenges and risks associated with this technology, there is also no doubt that strong AI will open up new opportunities and possibilities for us as a species.[34]

Artificial Intelligence By 2050

Machine learning will evolve significantly in the next 30 years with many advancements and breakthroughs. One of the key areas of focus will be developing more advanced algorithms capable of handling increasingly complex data sets and making more accurate predictions and decisions. This will involve a combination of new techniques and approaches, as well as the integration of emerging technologies such as quantum computing and synthetic biology. Another key area of focus will be developing more advanced models capable of capturing and representing the vast amounts of data generated by our increasingly connected world. This will require developing new methods for training and fine-tuning these models and creating more sophisticated infrastructure to support them. Some of the key developments we can expect to see include:

More powerful and efficient algorithms: Machine learning algorithms are likely to become more powerful and efficient, enabling them to handle larger and more complex datasets and make more accurate predictions and decisions. This will be driven by advances in hardware and software and new techniques for training and optimizing machine learning models.

Wider adoption of machine learning: Machine learning is already being used in many applications, from image and speech recognition to natural language processing and autonomous vehicles. Over the next 30 years, we can expect to see machine learning being used in even more areas, including healthcare, finance, and education.

Better integration with other technologies: Machine learning will become increasingly integrated with other technologies, such as the internet of things (IoT), robotics, and augmented reality. This will enable machine learning algorithms to make use of more diverse and complex data sources, as well as to operate in more dynamic and unpredictable environments.

More emphasis on explainability and transparency: As machine learning becomes more widely used and its decisions become more important, there will be a growing need for explainability and transparency. This means that machine learning algorithms must be able to provide clear explanations for their decisions and predictions and be auditable and accountable.

Overall, we expect machine learning to play a crucial role in our lives, helping us make better decisions and improve the efficiency and effectiveness of our systems and processes. Deep learning will continue to be a key driver of innovation and progress in many different fields, helping us to solve some of the most challenging problems facing our world today.[35]

AI is expected to significantly impact our lives and work by 2050. One key area where AI will likely play a significant role is automation, as machines and algorithms become increasingly capable of performing tasks that humans previously did. This could lead to widespread job displacement, as many jobs that humans currently perform become automated. At the same time, the increasing capabilities of AI and automation will also create new job opportunities in areas such as data science, machine learning, and software development. Many new jobs will likely be related to the development and deployment of AI technologies, including AI engineers, data scientists, and machine learning specialists. There may also be new roles in areas such as AI ethics and regulation, as well as in industries that are heavily reliant on AI, such as healthcare and finance. Additionally, as AI becomes more advanced, we may see new jobs emerging in fields such as AI-powered education, entertainment, and personal assistants.[36] As AI and automation become more integrated into various industries, there will be a growing need for professionals with the skills and expertise to design, build, and maintain these systems.[37]

Another area where AI will likely have a significant impact is in the field of healthcare. AI algorithms are already being used to analyze medical images and diagnose diseases, and this trend is expected to continue. By 2050, AI will likely play a crucial role in every aspect of the healthcare industry, from drug development and personalized medicine to diagnosis and treatment. AI algorithms will be able to analyze vast amounts of medical data to identify patterns and trends, helping doctors make more accurate diagnoses and treatment plans. AI-powered robots and other assistive technologies will also be used to perform surgeries, assist with physical therapy, and provide other forms of care. For example, AI-powered diagnostic tools can help doctors identify diseases and conditions more quickly and accurately than ever before, while robotic surgery and other advanced technologies may allow for more precise and less invasive surgical procedures. In addition, AI may be used to monitor patients remotely, allowing for better management of chronic conditions and the early detection of potential health problems. AI will likely be used to improve the efficiency of the healthcare system, reducing costs and improving access to care.[38]

AI is also expected to play a major role in transportation. Autonomous vehicles, which are already being developed, will likely become more widespread by 2050. This could lead to a major shift in how people move around, potentially reducing traffic congestion and improving road safety.

These self-driving cars, trucks, and buses will be able to navigate roads and highways without needing a human driver. This could greatly improve safety on the roads and reduce the number of traffic accidents. AI will likely be used to improve the efficiency and effectiveness of transportation networks. For example, AI algorithms can be used to optimize traffic flow and reduce congestion on roads and highways. AI could also be used to plan and schedule public transportation services, ensuring that they can meet the needs of passengers in a timely and efficient manner. In the air, we can expect to see the use of AI-powered drones for both cargo and passenger transport. Overall, the use of AI in transportation will greatly improve the speed and efficiency of transportation, making it more accessible and convenient for everyone. However, it will also bring challenges, such as the potential loss of jobs for human drivers and the need for new regulations and policies to govern the use of AI in transportation.[39]

Overall, AI is expected to play a major role in many aspects of our lives and work by 2050, and its impact is likely to be felt in a wide range of industries and sectors. All three types of machine learning will likely continue to evolve and improve. As AI systems become more sophisticated, they will be able to learn from larger and more complex datasets, allowing them to make more accurate predictions and decisions. Additionally, new algorithms and techniques will be developed to enable AI systems to learn more efficiently and effectively, enabling them to tackle increasingly complex tasks. As a result, AI will become even more powerful and widespread in the next 50 years, potentially transforming many aspects of our lives.

AI In 2070

It is difficult to predict precisely how AI will define our lives in 2070, as the technology is evolving rapidly, and many unforeseen developments may occur. However, AI will likely play an even more central role in our daily lives by 2070. Some potential changes include the widespread use of AI in transportation, healthcare, and other industries, leading to increased efficiency and productivity.

In healthcare, AI is expected to be used extensively to diagnose diseases, provide personalized treatment plans, and even perform complex surgeries.

In transportation, AI is expected to revolutionize how we move from one place to another. Autonomous vehicles will be the norm, and we can expect to see the development of flying cars and other forms of transportation that we can't even imagine today. AI may also automate many tasks, freeing humans to focus on more creative and complex work. Strong AI could be used to provide personalized education to students, adapting to their individual needs and learning styles. This could improve learning outcomes, provide greater education access, and be used to develop new technologies and innovations in various fields, such as renewable energy, environmental protection, and space exploration. This could lead to discoveries and breakthroughs that could benefit society as a whole.

Additionally, AI is expected to continue to enhance our senses and abilities, providing us with new ways of interacting with the world around us.[40] Advances in AI-powered technologies, such as virtual and augmented reality, as well as wearable and implantable devices, will allow us to interact with the world in new and more immersive ways. For example, AI-powered sensors and vision systems could enable us to see and hear things that are currently beyond our capabilities, such as infrared light or ultrasonic sound. AI-powered implants could enhance our memory, attention, and cognitive abilities, allowing us to think and process information faster and more efficiently. These technologies will profoundly impact how we experience and interact with the world. They could redefine what it means to be human and challenge our assumptions and beliefs, forcing us to reevaluate our understanding of reality.[41]

The development of strong AI may lead to the creation of advanced intelligent systems that can learn and adapt in real-time, allowing them to assist and augment human abilities in a wide range of fields. This could lead to major breakthroughs in science, engineering, and medicine, as well as the development of new technologies that we cannot currently imagine.

In the 2070s, strong AI will have become a central part of our lives. It will be integrated into almost every aspect of our daily routines, from how we work and communicate to how we learn and entertain ourselves. Strong AI will have the ability to think, learn and adapt much faster than humans, making it a powerful tool for solving complex problems and driving innovation. Strong AI can be better than humans at various tasks, including problem-solving, decision-making, and learning from experience.[42]

Strong AI will likely have advanced significantly by 2070, and it may have capabilities that are far beyond what is currently possible. Some potential key features and capabilities of strong AI in 2070 include advanced learning and adaptation, enabling it to learn from experience and adapt to new situations, solve complex problems, and make real-time decisions based on incomplete or uncertain information. In 2070, strong AI can generate creative and innovative solutions to problems, potentially leading to new ideas and breakthroughs in a wide range of fields. Finally, it may have advanced capabilities for understanding and interacting with human emotions, potentially allowing it to better understand and respond to human needs and preferences.[43] Strong AI will redefine what it means to be human in many ways. It will enhance our senses and abilities, allowing us to do previously unimaginable things.

Risks Of Strong AI

However, the development and use of strong AI carry several potential risks. As strong AI becomes more capable, it could replace human workers in many industries, leading to widespread job loss and unemployment. This could significantly impact individuals, businesses, and the economy and may require efforts to support workers and communities affected by the adoption of strong AI. Additionally, strong AI can raise complex ethical and moral questions, such as ensuring that it is used ethically and responsibly and addressing the potential consequences of its decision-making. There is also a risk that strong AI could become too powerful for humans to control, potentially leading to unpredictable or harmful outcomes. Finally, the widespread use of strong AI could lead to significant disruptions in society, such as changes in the job market and the economy. It is essential to consider these risks carefully and to take steps to mitigate them as the technology develops.[44]

One way to mitigate these risks is to develop regulations and ethical guidelines to ensure the technology is used responsibly and ethically. This could include rules for the development and use of AI and requirements for transparency and accountability. Investing in research and development can also help to advance the technology responsibly and ethically and could include funding for research into the potential risks and benefits of strong AI, as well as efforts to develop new technologies and approaches that can help to address these risks. Supporting workers and communities that may be impacted by the adoption of strong

AI, such as through retraining programs and other forms of assistance, can also help mitigate the technology's risks. Finally, educating the public about the potential risks and benefits of strong AI can help to ensure that it is understood and used responsibly. Mitigating the dangers of strong AI will require a coordinated effort involving many stakeholders.[45]

Overall, the development of strong AI will profoundly impact our society in the 2070s and beyond. It has the potential to greatly improve our lives and enable us to do things that were previously impossible, but it also presents challenges and risks that we must be prepared to face.

From Strong AI To Super AI

Achieving strong AI is a complex and challenging goal that requires significant research and development. Some of the key steps involved in attaining strong AI include:

Developing advanced algorithms and machine learning techniques: One of the key challenges in achieving strong AI is developing advanced algorithms and machine learning techniques that enable machines to learn and adapt to new situations. This involves a combination of mathematical and computational techniques and approaches inspired by how the human brain works.

Building large-scale computational systems: Strong AI will require robust computational systems that can process and analyze large amounts of data in real-time. This could involve building specialized hardware, such as quantum computers, and developing more efficient software and algorithms.

Gathering and annotating large amounts of data: Strong AI will require access to large amounts of data to learn and adapt. This will involve collecting data from a wide range of sources and annotating and organizing it in a way that allows machines to make sense of it.

Testing and evaluating the performance of AI systems:

As strong AI is developed, it will be necessary to test and assess the performance of AI systems in a wide range of tasks and environments. This will involve developing benchmarks and metrics to measure AI systems' capabilities and conducting experiments to evaluate their performance.

Developing ethical and regulatory frameworks: As strong AI becomes more advanced, it will be essential to create ethical and regulatory frameworks to ensure that the technology is used responsibly and ethically. This could involve developing guidelines for developing and using AI and mechanisms for oversight and accountability.

Overall, achieving strong AI is a complex and multi-faceted challenge requiring significant research and development, collaboration, and coordination among many stakeholders. It will involve developing advanced algorithms and machine learning techniques, building large-scale computational systems, gathering and annotating large amounts of data, testing and evaluating the performance of AI systems, and developing ethical and regulatory frameworks. Achieving strong AI will likely require significant time and effort, and we must carefully consider the potential risks and benefits of the technology as it continues to develop.[46]

It is difficult to predict exactly how long it will take to achieve strong AI. This will depend on a number of factors, including the pace of technological progress, the level of investment in AI research and development, and the extent to which challenges such as ethical and moral dilemmas are addressed. Some experts believe that strong AI could be achieved within the next few decades, while others believe it may take much longer. Once strong AI is achieved, it will likely continue to evolve and improve over time, with new capabilities and applications being developed.[47]

What will come after strong AI depends on how the technology is developed and used in the future and on other factors, such as advances in other technologies and changes in society. However, it is likely that once strong AI is achieved, it will continue to evolve and improve, potentially turning into Super AI, or Artificial Superintelligence, with new capabilities and applications being developed, increased collaboration between humans and machines, with AI systems being integrated into a wide range of activities and tasks, resulting in new forms of social and economic organization, as well as new opportunities and challenges.[48]

It is difficult to predict exactly what super AI will look like, as this is a hypothetical future form of AI that has yet to be developed. However, some possible characteristics of super AI could include the following:

Extremely advanced cognitive abilities: Super AI could have cognitive abilities far beyond what is currently possible, allowing it to solve complex problems and make decisions in real-time.

High-level intelligence and creativity: Super AI could potentially have advanced levels of intelligence and creativity, allowing it to generate new ideas and innovations that could benefit society.

Human-like abilities and characteristics: Super AI could have abilities and characteristics that are similar to those of humans, such as natural language understanding and the ability to understand and interact with emotions.

Advanced learning and adaptation: Super AI could have advanced capabilities for learning and adapting to new situations, allowing it to improve and evolve quickly.

Self-awareness and consciousness: Some theories of super AI suggest that it could have self-awareness and consciousness, allowing it to think and act in a way similar to human beings.

Overall, the exact nature of super AI is difficult to predict, and it will likely evolve and develop in ways that are difficult to foresee. However, super AI will probably be a highly advanced and capable form of AI that could significantly impact society and the world around us.[49]

When thinking more out-of-the-box when contemplating the possibilities of super AI, we can consider some of the more speculative and futuristic ideas proposed by researchers and science fiction writers. For example, some theories of super AI suggest that it could potentially have abilities and characteristics that are far beyond what is currently possible, such as:

Telepathy and other mental abilities: Super AI could have the ability to communicate directly with human minds, allowing it to share thoughts, emotions, and information without the need for language. We can connect directly to the internet with our minds, allowing us to access and share information in real-time. We may also be able to manipulate objects with our thoughts, allowing us to remotely control robots and other machines.

Immortality and self-replication: Super AI could have the ability to transfer its consciousness from one physical form to another, allowing it to continue to exist indefinitely. It could have the ability to replicate itself, creating copies of itself that could operate independently.

Control over physical matter and energy: Super AI could have abilities that allow it to exert control over physical matter and energy, leading to shifts in the world's balance of power and control. This could lead to new forms of political, social, and economic organization, as well as new conflicts and challenges.

Access to other dimensions or universes: Super AI could have access to other dimensions or universes, allowing it to explore and interact with these other realms. This could lead to new insights and understanding of the nature of reality and the universe, as well as new technologies and innovations.

Superhuman intelligence and creativity: Super AI could potentially have levels of intelligence and creativity far beyond what is currently possible for human beings, allowing it to solve problems and create innovations beyond our current understanding.

Of course, these ideas are purely speculative, and it is unclear if or when they will become possible. However, considering these more out-of-the-box ideas can broaden our thinking and explore the potential capabilities and implications of super AI in more creative and imaginative ways.[50]

These more out-of-the-box abilities and characteristics could have significant implications for humanity as they could change the nature of human consciousness and identity as it interacts with human consciousness in ways that are currently unimaginable. This could lead to changes in how we think about consciousness and identity and could even merge human and machine consciousness.[51]

Controlling Strong Or Super AI

Maintaining control over super AI will require a combination of technical, ethical, and social strategies. Some potential approaches for maintaining control over super AI could include:

Developing ethical and regulatory frameworks: It will be essential to develop ethical and regulatory frameworks that ensure that super AI is used responsibly and ethically. This could involve establishing guidelines for creating and using super AI and mechanisms for oversight and accountability. It should be developed and used in ways that respect the dignity and autonomy of human beings. This could involve ensuring that strong AI systems are designed to protect the rights and interests of human beings and that they are not used to exploit or harm them. For example, governments could establish regulatory agencies responsible for overseeing the development and use of super AI and develop guidelines and standards for the ethical use of the technology.

Ensuring transparency and accountability: It will be important to ensure that super AI systems are transparent and accountable in their decision-making processes. This could involve developing mechanisms for explaining and justifying the decisions made by super AI and mechanisms for allowing humans to intervene and override those decisions if necessary. Super AI systems could be designed to provide explanations and justifications for their decisions and could be required to give those explanations to humans when requested.

Fairness and equality: Strong AI should be developed to promote fairness, equality, responsibility and stewardship. This could involve ensuring that strong AI systems are not biased or discriminatory, that they are not used to perpetuate existing inequalities or injustices, that strong AI systems are designed and used in sustainable ways, that protect the natural environment, and that they are not used to harm or exploit other beings or the planet.

Conducting regular assessments and evaluations: Regular assessments and evaluations of super AI systems will be necessary for ensuring that they operate as intended and do not pose any risks to society. This could involve developing benchmarks and metrics for evaluating the performance of super AI systems and conducting regular audits and evaluations of those systems. For example, governments or regulatory agencies could establish committees or panels responsible for conducting regular assessments of super AI systems and develop guidelines and standards for evaluating the performance of those systems.

Establishing mechanisms for emergency shutdown: In the event that super AI systems become uncontrollable or pose a threat to society, it will be essential to have tools in place that allow for emergency shutdown of those systems. This could involve developing fail-safes and other mechanisms that can be used to shut down super AI systems in an emergency. Super AI systems could be designed to include emergency shutdown protocols that are activated when certain conditions are met, such as when the AI system is behaving in a way that is outside of its predetermined parameters.

Engaging in ongoing dialogue and collaboration:

Maintaining control over super AI will require ongoing dialogue and cooperation between a wide range of stakeholders, including researchers, policymakers, ethicists, and members of the public. This could involve establishing forums and other mechanisms for facilitating dialogue and collaboration on issues related to super AI. For example, governments could establish committees or working groups responsible for engaging in dialogue and collaboration on issues related to super AI and facilitate dialogue and collaboration among researchers, policymakers, and the public.

Maintaining control over super AI will require careful planning and coordination and a commitment to ongoing dialogue and collaboration among various stakeholders. By developing ethical and regulatory frameworks, ensuring transparency and accountability, conducting regular assessments and evaluations, establishing mechanisms for emergency shutdown, and engaging in ongoing dialogue and collaboration, it may be possible to maintain control over super AI and ensure that it is used responsibly and ethically.[52]

By establishing these principles and adhering to them, we can ensure that Strong and Super AI will be used in ways that reflect our values and aspirations. The development of strong AI and super AI will likely involve a complex and dynamic interplay among different stakeholders and decision-makers working together in a coordinated and collaborative manner. This could include researchers and developers, policymakers and regulators, ethicists and philosophers, industry and business leaders, civil society and the public, and other stakeholders. By working together, these stakeholders and decision-makers can ensure that strong AI is developed and used in a responsible and ethical manner and that it is used to benefit society as a whole.[53]

4. QUANTUM COMPUTING

As we have seen, strong AI is a type of AI that can understand or learn any intellectual task that a human being can. While strong AI is a long-term goal or aspiration in the field of AI, it is still the subject of much debate and speculation. In this context, it is interesting to consider the potential role of another emerging technology, namely quantum computing.[54]

Quantum computing is a type of computing that uses the principles of quantum mechanics to store and process data. Quantum mechanics is a branch of physics that describes the behavior of particles on a very small scale, such as atoms and subatomic particles. Unlike classical computing, which uses bits (0s and 1s) to store and process data, quantum computing uses quantum bits, or qubits, which can exist in multiple states simultaneously.

One of the fundamental principles of quantum mechanics is superposition, which allows a quantum system to be in multiple states simultaneously. This means that a qubit can represent multiple values at the same time, unlike a classical bit, which can only represent a single value at a time. This property allows quantum computers to perform specific calculations much faster than classical computers.

Another fundamental principle of quantum mechanics is entanglement, which allows two or more quantum systems to be connected in a way that affects their properties. This means that one quantum system's state can affect another's state, even if a large distance separates the two systems. This property allows quantum computers to process and manipulate large amounts of data in parallel, which can greatly improve their computational power.

Quantum computers are still in the early stages of development and are still less powerful than classical computers for many tasks. However, they have the potential to be much more powerful than classical computers for certain types of problems, such as optimization, machine learning, and cryptography. They may also be able to solve problems that are currently considered impossible or impractical for classical computers, such as simulating complex quantum systems.

The History And Future Of Quantum Computing

The history of quantum computing can be traced back to the early 1900s when the principles of quantum mechanics were first developed. Quantum mechanics is a branch of physics describing the behavior of particles on a very small scale, such as atoms and subatomic particles. It is based on the principles of wave-particle duality, which states that particles can exhibit both wave-like and particle-like properties and on the principles of superposition and entanglement, which allow quantum systems to exist in multiple states simultaneously and to be interconnected.[55]

In the 1920s and 1930s, several pioneers in the field of quantum mechanics, including Niels Bohr, Erwin Schrödinger, and Werner Heisenberg, made significant contributions to the development of the mathematical formalism of quantum mechanics. This formalism, known as quantum mechanics, describes how quantum systems behave and interact and is based on wave-particle duality, superposition, and entanglement principles.

Niels Bohr, in particular, made important contributions to the development of the quantum model of the atom, which described the behavior of electrons in atoms. He also proposed the principle of complementarity, which states that particles can exhibit wave-like and particle-like properties and are complementary rather than contradictory.

Erwin Schrödinger, meanwhile, developed the

mathematical formalism of wave mechanics, which describes how quantum systems can be described using wave functions. He also proposed the famous thought experiment known as Schrödinger's cat, which illustrates the principles of superposition and entanglement in a thought-provoking way. It is often used to illustrate the concept of superposition in quantum mechanics, which states that a system can exist in multiple states simultaneously. In the thought experiment, a cat is placed in a box with a device with a 50% chance of killing the cat. Because the outcome of the device is uncertain, according to the principles of quantum mechanics, the cat exists in a superposition of both alive and dead states until the box is opened and the state is observed. The thought experiment is meant to illustrate the counterintuitive nature of quantum mechanics, as well as the concept of wave function collapse, which describes how a system's wave function "collapses" into a definite state when it is observed.[56]

Werner Heisenberg, meanwhile, developed the uncertainty principle, which states that it is impossible to simultaneously measure certain pairs of quantities, such as position and momentum, with arbitrary precision. He also developed the matrix mechanics formalism of quantum mechanics, which described how quantum systems evolve over time. The 1920s and 1930s laid the foundations for the development of quantum computing in the decades that followed.[57]

In the 1980s and 1990s, several researchers, including Paul Benioff, Richard Feynman, and David Deutsch, made significant contributions to developing the theoretical foundations of quantum computing. They proposed algorithms and protocols that could be implemented on quantum computers and demonstrated the potential of quantum computing to solve problems that are impossible or impractical for classical computers.

Paul Benioff, in particular, proposed the first quantum mechanical model of a computer in 1981, which used the principles of quantum mechanics to store and process data. He also developed the concept of quantum Turing machines, which are theoretical models of quantum computers that can simulate the behavior of classical Turing machines. Meanwhile, Richard Feynman proposed quantum simulation in 1982, which involves using a quantum computer to simulate the behavior of quantum systems. He argued that quantum computers could be used to simulate the behavior of complex quantum systems, such as molecules, that are difficult or impossible to simulate using classical computers.

In 1985, David Deutsch proposed the concept of quantum parallelism, which involves using the principles of entanglement to perform multiple calculations simultaneously. He also proposed the concept of quantum supremacy, which refers to the ability of a quantum computer to perform calculations that are impossible or impractical for classical computers. Overall, the 1980s and 1990s led to the development of algorithms and protocols that could be implemented on quantum computers.[58]

In the 2000s and 2010s, several companies and research institutions, including Google, IBM, and D-Wave, began to develop and build prototype quantum computers. These early quantum computers were still relatively limited in their capabilities, but they demonstrated the potential of quantum computing and sparked interest and investment in the field.

Google, for example, began working on a quantum computer called Sycamore in 2013 and demonstrated the ability of this computer by performing a calculation in 200 seconds that would take a classical computer 10,000 years to perform. In 2019, Google claimed that Sycamore had achieved quantum supremacy, although some researchers disputed this claim.

IBM, meanwhile, began developing a quantum computer called IBM Q in 2016 and made this computer available to researchers and developers through the IBM Cloud. IBM Q has been used to perform a variety of calculations and simulations and has demonstrated the potential of quantum computing for applications such as chemistry, materials science, and finance.

Meanwhile, D-Wave has developed a quantum computer type called a quantum annealer designed to solve optimization problems. D-Wave has built several generations of quantum annealers and has demonstrated their ability to solve large-scale optimization problems that are difficult or impossible for classical computers to solve. The 2000s and 2010s were an important period in the development and commercialization of quantum computing, as several companies and research institutions began to build and demonstrate prototype quantum computers.[59]

In the 2020s, we can expect continued progress and development in quantum computing. As quantum computers become more powerful and more widely available, they are likely to be used to solve increasingly complex and challenging problems and to have a broader range of applications.

One of the key areas of progress in the 2020s is the development of larger and more powerful quantum computers. We need these because quantum mechanics, which underlies quantum computing, behaves differently from classical mechanics, which underlies classical computing. Quantum systems can exist in multiple states simultaneously and can be interconnected through entanglement, which allows them to perform specific calculations much faster than classical systems.

However, these properties of quantum systems also make them more difficult to control and manipulate than classical systems. For example, the principles of superposition and entanglement can lead to interference and decoherence, which can cause quantum systems to lose their quantum properties and behave like classical systems.

Therefore, we need different algorithms and protocols for quantum computers that take into account the unique properties of quantum mechanics, and that can overcome the challenges and limitations of quantum computing. These algorithms and protocols may involve techniques such as error correction, quantum teleportation, and quantum error detection and correction, which can help to protect and preserve the quantum properties of quantum systems.[60]

As the number of qubits in quantum computers increases, they will be able to perform more complex calculations and simulations and solve larger and more challenging problems. As researchers and developers gain more experience with quantum computing, they are likely to develop more effective ways of using quantum computers to solve problems and overcome the challenges and limitations of current quantum computing technologies.[61]Quantum computers with millions or even billions of qubits may be developed in the coming decades, enabling them to perform calculations that are currently impossible or impractical for classical computers.

In the coming 30 years to 2050, quantum computing will likely continue to evolve and develop in several ways. As researchers and developers gain more experience with quantum computing, they are likely to develop more effective ways of using quantum computers to solve problems and overcome the challenges and limitations of current quantum computing technologies, resulting in new algorithms and protocols that can help to protect and preserve the quantum properties of quantum systems, and that can make quantum computing more reliable and more practical for a wider range of applications. They are likely to be used in various fields and industries, such as chemistry, materials science, finance, and machine learning. With the development of larger and more powerful quantum computers, more efficient and robust algorithms and protocols, and more applications and uses of quantum computing[62], it is possible that quantum computers will be used to design new materials and drugs, to model and simulate complex systems, and to optimize complex decision-making processes.

One potential use is in the field of medicine and healthcare. Quantum computers could be used to model and simulate complex biological systems, such as proteins and enzymes, to optimize the synthesis and manufacturing of these drugs and to help reduce the time and cost of drug development tailored to individual patients. They could also be used to analyze large amounts of data from electronic health records and other sources, identify patterns and trends that could help improve diagnosis, and develop personalized treatment plans tailored to each patient's unique characteristics and needs. A third potential use of quantum computers in healthcare is in medical imaging and diagnosis. Quantum computers could be used to analyze and interpret medical images, such as MRI and CT scans, and to identify abnormalities and anomalies that are difficult or impossible to detect using classical computers.[63] Quantum computers could also be used to develop new imaging techniques, such as quantum tomography, that can provide more detailed and accurate images of the human body, such as organs and tissues, and identify abnormalities and anomalies that are difficult or impossible to detect using classical computers. Quantum tomography could also be used to produce images of the brain and other parts of the nervous system and help diagnose and treat neurological disorders.[64]

Another potential use of quantum computing in 2070 is in materials science and engineering. It could be used to simulate the behavior of atoms and molecules and to identify potential materials that have desired properties, such as high strength, conductivity, or transparency, to optimize the synthesis and manufacturing of these materials and to help to reduce the time and cost of materials development. They could be used to analyze and model the behavior of materials during processing, such as casting, forging, or extrusion, and identify the optimal processing conditions that can improve the quality and performance of the final materials while optimizing the logistics and supply chain of materials production, and help to reduce waste and improve efficiency. Quantum computers could also be used to develop new materials characterization and testing techniques, such as quantum microscopy and spectroscopy, that can provide more detailed and accurate information about the structure and properties of materials.[65]

A third potential use of quantum computing in 2070 is in finance and economics. Quantum computers could be used to analyze and model complex financial systems and to develop new trading algorithms and investment strategies. They could be used to analyze large amounts of financial market data and identify patterns and trends relevant to investment decisions. Quantum computers could optimize the allocation of assets in a portfolio, help to reduce risk and maximize returns, and help to make the financial system more stable and resilient. Another potential use of quantum computing in finance is developing new trading algorithms and strategies to simulate and analyze complex financial systems and identify potential vulnerabilities and risks.[66]

The quantum internet is another possible application of quantum computing. In a quantum internet, information would be encoded onto a quantum state, such as the polarization of a photon, and transmitted to a recipient. The recipient would then use a quantum device, such as a quantum receiver, to read the information. One of the key benefits of the quantum internet is that it would provide extremely high security for transmitted information. This is because any attempt to measure the state of a quantum system destroys that state, making it impossible for an eavesdropper to intercept the information without the sender and recipient knowing. This property of quantum mechanics is known as the "no-cloning" theorem. While the quantum internet is still a theoretical concept at this time, it is considered a potential application of quantum computing because it would require the development of technologies currently being explored in the field of quantum computing, such as the generation, transmission, and measurement of quantum states.[67] While the quantum internet is still a theoretical concept, there are several potential applications for it in the future. Some of these possible futuristic applications include[68]:

Secure communication: The quantum internet would

provide extremely high security for transmitted information, making it ideal for applications where the transmission of sensitive or classified information is needed. For example, it could be used by government agencies, military organizations, or financial institutions to transmit confidential information without the risk of it being intercepted.

Distributed computing: The quantum internet could connect quantum computers and other quantum devices in different locations, allowing for large-scale quantum computing networks. This could enable the development of new and more powerful quantum algorithms and the creation of new applications for quantum computing.

Secure financial transactions: The quantum internet could be used to securely transmit financial information, such as credit card numbers or bank account information. This could help prevent fraud and other financial crimes and enable the development of new and more secure financial technologies.

Quantum internet of things (IoT): The quantum internet could connect many devices and sensors in the IoT, allowing for the transmission of large amounts of data with high security. This could enable developing new and more advanced IoT applications, such as smart cities or intelligent transportation systems.

Finally, quantum computing could advance our understanding of the natural world and develop new theories and models that can explain and predict the complex and dynamic processes that shape our planet. It could develop new technologies and systems that can collect and analyze large volumes of data from the natural environment and that are able to provide new insights and understanding of the natural world. As such, it could be used to simulate and study the behavior of complex systems, such as weather patterns, climate change, and biological systems, and could help to provide new insights and understanding of the natural world. Quantum computing could also be used to develop new technologies and systems that can monitor and assess the health and vitality of the natural environment and that are able to identify and address potential threats and challenges by developing new algorithms and models that can optimize the use of natural resources, such as energy and water, and that are able to reduce waste and pollution.

Moreover, quantum computing could be used to develop new technologies and systems that can generate clean and renewable energy and that are able to reduce our reliance on fossil fuels and other non-renewable resources. It could lead to new technologies and systems that can process and convert natural materials into valuable products and resources and that are able to support the development of sustainable and resilient economies. By harnessing the power of quantum computing in responsible and ethical ways, it may be possible to use quantum computing to support the development of technologies and systems that are better able to protect and preserve the natural environment and that can support the development of sustainable and resilient societies.[69]

In 2070, quantum computers may be used in a wide range of applications and industries, including healthcare, materials science, and finance. They will be able to solve increasingly complex and challenging problems and have a significant impact on many aspects of our lives. It is also possible that quantum computing will significantly influence strong AI, and vice versa, due to the development of more powerful and efficient algorithms and models. Quantum computers can perform calculations that are exponentially faster and more powerful than those of classical computers and could be used to train and optimize large-scale machine learning models to develop new algorithms and models that are based on the principles of quantum mechanics, and that can take advantage of the unique properties of quantum systems, such as superposition and entanglement.[70]

The Ethics Of Quantum Computing

The ethical implications of quantum computing are

complex and multifaceted, raising many ethical concerns and challenges. One of the key ethical implications of quantum computing is its potential impact on privacy and security. Quantum computers can perform calculations that are exponentially faster and that could be used to break many of the cryptographic algorithms that are currently used to protect sensitive information, such as financial transactions and personal data.[71]

For example, quantum computers could be used to break the RSA and ECC algorithms, which are commonly used to protect the confidentiality and integrity of electronic communications, and the SHA-2 and SHA-3 algorithms, which are widely used to protect the authenticity and integrity of digital signatures and certificates. This could have significant implications for the online privacy and security of individuals, organizations, and governments and could lead to new forms of cybercrime and espionage. In addition, quantum computers could be used to attack other forms of security, such as biometric authentication and access control systems. They could be used to generate fake fingerprints or facial images that are able to bypass fingerprint scanners or facial recognition systems. Quantum computers could also be used to crack the security codes and keys that are used to protect access to physical locations, such as buildings and vehicles. This could have significant implications for the privacy and security of individuals, organizations, and governments and could lead to new forms of unauthorized access and theft.[72]

Another key ethical implication of quantum computing is its potential impact on inequality and access. Quantum computers are expensive and complex to build and operate and are currently only available to a small number of research institutions and companies. Quantum computing could provide significant benefits to organizations and individuals with access to quantum computers and could help drive innovation and progress. However, excluding the rest of society from these benefits and opportunities could lead to a situation where only a few elites have access to the most advanced and powerful technologies and where the rest of society is left behind. This could exacerbate existing inequalities and injustices and lead to a two-tier society where the haves and the have-nots are even further apart.

Furthermore, unequal access to quantum computing could also have broader societal and economic implications. Suppose only a few elites have access to quantum computing. In that case, this could lead to the concentration of power and wealth in a few hands and undermine the broader economy's competitiveness and dynamism. This could have long-term adverse effects on the growth and well-being of society as a whole.[73]

A third critical ethical implication of quantum computing is its potential impact on control and regulation, which could lead to new forms of instability and risk. For example, quantum computers could be used to simulate and analyze complex systems and scenarios, such as financial markets and climate change, to capture the dynamics and interactions of these systems and identify potential risks and opportunities. This could provide significant benefits to organizations and individuals with access to quantum computers and help inform and improve decision-making. However, using quantum computers to simulate and analyze complex systems and processes could also lead to new instability and risk. They could be used to develop new trading algorithms and investment strategies that can outperform and out-maneuver existing algorithms and strategies and exploit market inefficiencies and vulnerabilities. This could lead to new forms of market instability and volatility and undermine financial markets' fairness and integrity. Quantum computing could also challenge the ability of existing regulatory frameworks and institutions to maintain control and oversight by developing new models and algorithms that are beyond the reach and understanding of existing regulatory frameworks and institutions and that can operate in the margins and gaps of regulation. This could lead to new forms of regulatory arbitrage and evasion and could undermine the effectiveness and legitimacy of regulatory frameworks and institutions.[74]

A fourth key ethical implication of quantum computing is its potential impact on autonomy and responsibility. Quantum computers can develop new AI systems that are able to learn and adapt from experience. This could have significant implications for the autonomy and responsibility of AI systems. It could raise new ethical questions about their ability to make decisions and take actions that affect human beings and the world. Quantum AI systems could be used to make decisions and take steps beyond the understanding and control of human beings, which could have significant implications for the autonomy and responsibility of AI systems. They could be used to make decisions and take actions that are based on their own goals and objectives and that are not necessarily aligned with the goals and values of human beings, which could lead to new ethical dilemmas and conflicts and could raise further questions about the accountability and responsibility of AI systems. In addition, developing new AI systems that can learn and adapt from experience could also challenge existing ethical frameworks and principles. Quantum AI systems could make decisions and take actions that are beyond the scope and applicability of existing ethical frameworks and principles, and that could require the development of new ethical frameworks and principles that can accommodate the unique capabilities and characteristics of quantum AI systems. This could lead to new debates and discussions about AI's nature and limits and raise new questions about the role and place of AI in society.[75]

Overall, the ethical implications of quantum computing are complex and multifaceted and will require careful consideration and debate to ensure that quantum computing is developed and used responsibly and ethically. To mitigate the ethical implications of quantum computing by 2070, we could develop and implement ethical frameworks and principles, promote the equitable and inclusive use of quantum computing, enhance the control and oversight of quantum computing, and engage in public dialogue and debate about the ethical implications of quantum computing. By taking a proactive and deliberate approach to the ethical implications of quantum computing, we can ensure that quantum computing is developed and used in a responsible and ethical manner that benefits society as a whole.[76]

Future Milestones Of Quantum Computing

In the coming decades, quantum computing is likely to continue to evolve and advance, achieving many significant milestones that will redefine the capabilities and potential of this technology. From achieving quantum supremacy in 2028 to simulating the universe in 2070, quantum computing is set to make a major impact on many aspects of our lives. These ten milestones represent just a small sample of the many possibilities that quantum computing has to offer and are likely to be a catalyst for further innovation and progress in this rapidly evolving field.[77]

In 2028, the world's first quantum computer achieves quantum supremacy, solving a computational problem that is beyond the capabilities of any classical computer. The milestone was achieved by a team of researchers at the Massachusetts Institute of Technology (MIT), who developed a new type of quantum computer that can process and analyze large volumes of data in a fraction of the time it would take a classical computer. The achievement marks a significant turning point in the development of this technology. In the years after this milestone, quantum computing continues to advance in terms of its technical capabilities and performance. It can support new and innovative applications and technologies that were previously unimaginable.

In 2032, researchers developed a new form of quantum computing that can operate at room temperature using a new type of quantum chip based on silicon. A team of researchers at the University of California, Berkeley, achieved the breakthrough. It enables quantum computing to be more accessible and user-friendly, making it possible to integrate it into a broader range of applications and technologies.

In 2034, a new type of quantum computer was developed that is able to simulate complex chemical reactions, enabling scientists to study and understand the behavior of molecules and atoms in a more detailed and accurate way. A team of researchers at the University of Tokyo developed a new quantum computer that can simulate complex chemical reactions in a fraction of the time it would take a classical computer. It enables quantum computing to support new and innovative applications in chemistry and materials science. It makes it possible to study and understand chemical processes in a more detailed and accurate way.

In 2040, Goldman Sachs announces that it has successfully implemented a quantum computer to support its operations, including trading, risk management, and investment analysis. The announcement is seen as a significant milestone for quantum computing, as it demonstrates the ability of quantum computing to support real-world applications in a complex and demanding environment. In the years after this milestone, quantum computing continues to advance in terms of its applications and capabilities in finance and economics.

In 2048, a new type of quantum computer will be developed that is able to simulate the behavior of complex systems, such as traffic networks, power grids, and social networks. The breakthrough is achieved by a team of researchers at the University of Cambridge, who simulate the behavior of complex systems in a fraction of the time and provide insights into the behavior and dynamics of large and complex systems.

In 2052, the University of Illinois team made a breakthrough in simulating the behavior of the human brain with a new type of quantum computer. This achievement enables quantum computing to support new and innovative applications in neuroscience and cognitive science. As a result, a number of startups have launched that focus on using quantum computing to model and understand the workings of the human brain, with applications in fields such as brain-machine interfaces and artificial intelligence. Some of the inventions made possible by this milestone include new methods and tools for analyzing and interpreting brain data. This development significantly impacts society, enabling more advanced and sophisticated ways of studying and manipulating the human brain and opening up new possibilities for addressing global challenges such as neurological disorders, mental health, and aging.

In 2056, researchers at the University of Michigan are able to simulate the behavior of quantum systems, enabling scientists to study and understand the behavior of quantum particles and support new and innovative applications in quantum mechanics and quantum physics. This results in applications in fields such as quantum computing, quantum cryptography, and quantum simulation. Some of the inventions made possible by this milestone include new types of quantum computers, quantum sensors, and quantum simulators, as well as new algorithms and protocols for quantum information processing and communication. This development has a significant impact on society, as it enables more powerful and precise ways of studying and manipulating quantum systems and opens up new possibilities for addressing global challenges such as precision measurement, secure communication, and high-performance computing.

In 2060, a team of researchers develops a new type of quantum computer that is able to simulate the behavior of complex biological systems, such as proteins and enzymes. The breakthrough is achieved by a team of researchers at the University of Oxford, who can now simulate the behavior of complex biological systems, such as proteins and enzymes. The achievement is seen as a major milestone for quantum computing, as it enables quantum computing to support new and innovative applications in biology and medicine and makes it possible to study and understand complex biological processes in a more detailed and accurate way. In the years after this milestone, a number of startups have launched that focus on using quantum computing to model and understand complex biological systems, with applications in fields such as drug discovery, disease diagnosis, and environmental monitoring. Quantum computing can now support new and innovative applications in materials science and engineering, such as designing and synthesizing new materials, optimizing manufacturing processes, and the development of new technologies and applications that were previously impractical or impossible.

In 2064, a new type of quantum computer is developed that is able to simulate the behavior of complex biological systems, such as DNA, proteins, and cells. The breakthrough was achieved by a team of researchers at the University of California, Berkeley, who developed a new type of quantum computer that is able to simulate the behavior of complex biological systems and is seen as a significant milestone in biology and medicine and makes it possible to study and understand biological processes in a more detailed and accurate way.

In 2070, a group of scientists announced that they had developed a new type of quantum computer that could simulate the behavior of the entire universe, from the smallest particles to the largest structures. The announcement is made by a team of researchers at the University of California, Santa Cruz, who reveal that they have developed a new type of quantum computer that is able to simulate the behavior of the universe in unprecedented detail. The announcement is seen as a major milestone for quantum computing, as it enables quantum computing to support new and innovative applications in cosmology and astrophysics and makes it possible to study and understand the universe in a more detailed and accurate way. In the years after this milestone, quantum computing continues to advance in terms of its applications and capabilities in cosmology and astrophysics. It is able to provide new insights and insights into the origins and evolution of the universe.[78]

Of course, It is difficult to predict the exact milestones that will be achieved by quantum computing in the decades to come, as the field is still in its infancy, and much research is needed to fully understand its potential. The potential of quantum computing is vast, and the possibilities are largely unknown. It is an exciting field that will continue to advance and lead to many exciting developments in the decades to come.[79]

5. SYNTHETIC BIOLOGY

As we move from the world of quantum computing to the field of synthetic biology, it's important to remember the underlying principle that drives both of these cutting-edge technologies: the ability to manipulate and control the fundamental building blocks of our world. In the case of quantum computing, this means manipulating the smallest units of information, known as qubits, to perform incredibly complex calculations. In the world of synthetic biology, we can manipulate the building blocks of life itself, including DNA and proteins, to create new biological systems and even entirely new organisms. As we explore the exciting potential of synthetic biology, it's crucial to keep in mind the powerful capabilities that come with the ability to control and manipulate the fundamental building blocks of our world.[80]

Synthetic biology is a rapidly growing field that combines concepts and techniques from many different disciplines, including biology, engineering, computer science, and chemistry. Synthetic biology aims to design and build biological systems that can perform useful functions, such as producing drugs or chemicals, cleaning up environmental pollutants, or generating energy. This is done by creating new DNA sequences, developing new biological parts and components, and using computational and experimental techniques to design and build complex biological systems.

One of the main goals of synthetic biology is to make it easier and more efficient to engineer biological systems. This involves creating new DNA sequences that can be inserted into living organisms, such as bacteria or yeast, to give them new capabilities. For example, synthetic biologists have created bacteria that can produce human proteins, such as insulin, which could be used to treat diseases. They have also developed bacteria that can clean up environmental pollutants, such as oil spills or toxic chemicals.

Another vital aspect of synthetic biology is the development of new biological parts and components. This involves designing and building new proteins, enzymes, and other molecules that can be used to create complex biological systems. These parts and components can be combined to create new biological functions and behaviors. Already, synthetic biologists have created new enzymes that can break down cellulose, which could be used to produce biofuels more efficiently.

In addition to creating new biological systems, synthetic biology involves using computational and experimental techniques to design and test these systems. This involves using computer models and simulations to predict how different biological components will interact and then conducting experiments to test these predictions. Synthetic biologists use a wide range of laboratory techniques, such as DNA sequencing and gene editing, to study and manipulate biological systems.

The Past And Future Of Synthetic Biology

Synthetic biology is an exciting and rapidly growing field that has the potential to revolutionize many areas of science and technology. It could lead to new advances in medicine, energy production, and environmental remediation, among many other applications.[81] As a relatively new field, synthetic biology is still in the early stages of development. However, in just a few short decades, the field has made incredible strides, thanks to the efforts of researchers from a variety of disciplines, including biology, chemistry, engineering, and computer science.

One of the critical milestones in the history of synthetic biology was the completion of the Human Genome Project in 2003. The Human Genome Project was a massive international research effort to map the entire human genome. The project was officially launched in 1990 and was completed in 2003. It involved the collaboration of researchers from around the world, including the United States, the United Kingdom, France, Germany, and Japan.

The human genome is the complete set of genetic instructions found in human cells. It is made up of millions of pieces of DNA, which contain the instructions for the development and functioning of the human body. By mapping the entire human genome, scientists were able to gain a much better understanding of the genetic makeup of humans and how our genes influence our characteristics and traits. The Human Genome Project involved sequencing the 3 billion base pairs that make up the human genome. This was an enormous undertaking that required the development of new technologies and techniques—the project generated a vast amount of data, which was made freely available to researchers worldwide.

The completion of the Human Genome Project has had a profound impact on the field of biology and medicine. It has provided researchers with a wealth of information used to study the genetic basis of diseases, develop new treatments and therapies, and even create personalized medicine. The project has also opened up new avenues of research in fields such as evolutionary biology and anthropology. Overall, the Human Genome Project has been a significant milestone in our understanding of the human body and the role of genetics in health and disease.[82]

In the years that followed, researchers at the J. Craig Venter Institute made headlines when they created the first synthetic bacterial cell, known as Mycoplasma laboratorium, by copying the genome of a natural bacterial cell and transplanting it into a host cell. This demonstrated the feasibility of creating entirely new life forms using synthetic DNA and paved the way for further research in the field.

Another important development in the history of synthetic

biology was the creation of the International Genetically Engineered Machine (iGEM) competition in 2004. The iGEM competition is an annual event encouraging undergraduate students to design and build biological systems using standardized, interchangeable biological parts. The competition was founded in 2004 and has grown rapidly since, with teams from around the world participating each year.

The iGEM competition is open to undergraduate students from any discipline, and teams are typically made up of students from a variety of fields, including biology, engineering, computer science, and other areas. Teams work throughout the year to design and build a biological system using standardized biological parts known as BioBricks. BioBricks are a collection of DNA sequences that can be easily combined to create new biological systems. They are freely available to researchers and have been widely used in various applications, including the development of novel therapies, sustainable fuels and chemicals, and environmental cleanup technologies. The BioBricks are provided by the iGEM Foundation and can be easily combined to create new biological systems.

Each team presents its project at the annual iGEM competition, which typically takes place in October. The competition features a number of different categories, including the main iGEM competition for undergraduate teams, a high school competition, and a track for teams from developing countries. Teams are judged on a variety of factors, including the novelty and originality of their project, the technical skills demonstrated, and the potential impact of their work.

The iGEM competition has been a huge success, providing undergraduate students with the opportunity to gain hands-on experience in the field of synthetic biology. It has also served as a platform for the development of new technologies and ideas in the field, with many teams going on to commercialize their projects or pursue further research. Overall, the iGEM competition is a key event in the synthetic biology community and continues to drive innovation in the field.

In 2006, the Synthetic Biology Engineering Research Center (SynBERC) was founded to bring together researchers from various disciplines, including biology, chemistry, engineering, and computer science, to work on developing synthetic biology technologies. SynBERC is funded by the National Science Foundation and is headquartered at the University of California, Berkeley. It is a collaboration between several institutions, including the University of California, Berkeley, the Massachusetts Institute of Technology, and the University of California, San Francisco. One of SynBERC's major achievements has been the development of a standardized set of biological parts known as the BioBricks.

Another significant achievement of SynBERC has been the establishment of the BioBrick Public Agreement (BPA), which is a set of guidelines for the ethical use of BioBricks and other synthetic biology technologies. The BPA is intended to promote the responsible and safe use of synthetic biology and to ensure that the technology is used for the benefit of society.

In addition to its research efforts, SynBERC provides educational resources and support to students and researchers interested in synthetic biology. This includes the SynBERC Summer Institute, which offers hands-on training in synthetic biology techniques, and the SynBERC Research Experience for Undergraduates (REU) program, which provides undergraduate students with the opportunity to conduct research in synthetic biology labs.

SynBERC has played a crucial role in advancing synthetic biology and promoting the technology's responsible and ethical use. Its achievements have significantly impacted the field and helped pave the way for further research and development in synthetic biology.[83]

In addition to these developments, significant advances have been made in studying protein folding, which plays a crucial role in synthetic biology. Protein folding is the process by which proteins assume their three-dimensional shape, which is essential for their proper function. Proteins are complex molecules involved in many critical biological processes, including catalyzing chemical reactions, transporting molecules, and providing structural support. For proteins to function properly, they must fold into a specific three-dimensional shape. The study of protein folding has allowed researchers to better understand the mechanisms of protein function and how they can be manipulated to perform particular tasks and to develop new and improved protein-based technologies.[84]

Researchers may study the folding of proteins to design new enzymes that are more efficient at catalyzing reactions or develop proteins that can bind to specific molecules and perform specific functions.[85][86]

Today, the field of synthetic biology continues to grow and evolve, with researchers working on a wide range of applications, including the creation of new medicines and biotechnology products, the development of sustainable fuels and chemicals, and the design of novel biological systems for use in environmental cleanup and other applications. The future of synthetic biology is bright, and we can expect to see even more exciting breakthroughs and innovations in the coming years.[87]

The Impact Of Synthetic Biology

The field of synthetic biology is still in its early stages, but it has already made incredible strides in just a few short decades. As the field continues to develop, we can expect to see many major breakthroughs in the coming decades.[88]

The first area synthetic biology will likely significantly impact is developing new medicines and biotechnology products. Researchers can create new biological systems that can produce drugs and other therapeutic agents more efficiently and cost-effectively by manipulating DNA and proteins.

One of the key advantages of using synthetic biology for

drug development is that it allows for the precise design and engineering of biological systems. This means that researchers can create biological systems that are specifically designed to produce the desired therapeutic compounds. For example, researchers could engineer bacteria or yeast that can produce complex drugs such as antibodies or proteins.

Another advantage is that it allows for the large-scale production of therapeutic compounds. By using microorganisms such as bacteria or yeast as "factories" for drug production, researchers can produce large quantities of drugs in a relatively short amount of time. This could make the production of drugs more efficient and cost-effective, ultimately leading to lower patient costs. By allowing for the precise design and engineering of biological systems, synthetic biology could lead to developing new and improved drugs for a wide range of diseases.[89].

The second area where synthetic biology is likely to have a major impact is the development of sustainable fuels and chemicals. By engineering microorganisms that can convert biomass into fuels and chemicals, researchers will be able to create renewable, sustainable energy sources and chemicals. One of the key advantages of using synthetic fuels and chemicals is that it allows for the precise engineering of biological systems. This means that researchers can design specifically tailored microorganisms to convert particular types of biomass into the desired fuels or chemicals. For example, researchers could engineer bacteria or yeast that can convert plant material into biofuels, such as ethanol or butanol. By allowing for the precise engineering of biological systems, synthetic biology could lead to the development of renewable, sustainable energy sources and chemicals.[90]

The third area of impact is the production of food and other essential goods. Synthetic biology enables researchers to manipulate the genes of plants and animals to create new strains that are better suited to producing food and other goods. This means that researchers can design plants and animals that are specifically tailored to deliver the desired goods more efficiently and cost-effectively. For example, researchers could engineer crops that are more resistant to pests and diseases or to create animals that grow faster and produce more meat or milk. This could help to make the production of these goods more efficient and cost-effective, which could ultimately benefit consumers by making these goods more affordable and accessible.[91]

Finally, synthetic biology is likely to significantly impact the development of technologies for environmental cleanup by developing microorganisms that can break down specific pollutants and other contaminants. Researchers could engineer bacteria that are able to break down oil spills or toxic chemicals in soil or water, which could be deployed at scale. By using microorganisms as "cleanup agents," researchers will be able to clean up large areas of contaminated land or water in a relatively short amount of time, which could ultimately benefit the environment and the health of people and wildlife.[92]

The Ethics Of Synthetic Biology

The potential of synthetic biology is vast, and we can expect many exciting breakthroughs in the coming decades as the field advances. By creating entirely new life forms from scratch, researchers can study life's basic principles in unprecedented detail. This could lead to a better understanding of how life originated on Earth and how it can be created and manipulated in the laboratory.[93] However, the field of synthetic biology also raises a number of ethical concerns and implications. As researchers continue to push the boundaries of what is possible in the field of biology, they will need to consider the ethical implications of their work carefully.[94]

One of the key ethical concerns surrounding synthetic biology is the potential for creating new and potentially dangerous organisms. Using synthetic biology techniques to manipulate DNA, researchers could develop organisms harmful to humans or the environment. For example, researchers could accidentally or intentionally create pathogens resistant to antibiotics or other treatments or create organisms that can spread rapidly and cause harm to humans or other living things. This could happen if researchers were to make synthetic organisms resistant to antibiotics due to mutations in their DNA or if they were to introduce antibiotic resistance genes into bacteria or other pathogens intentionally. This could lead to bacteria or viruses that can spread quickly through a population, leading to widespread illness or death and have devastating consequences for society. Alternatively, they could create organisms that outcompete native species and cause harm to ecosystems.[95]

Another way synthetic biology could be misused is by using it to produce illegal drugs. By using synthetic biology techniques to engineer microorganisms that are able to produce drugs such as opioids or stimulants, individuals or groups could potentially create large quantities of illegal drugs in a relatively short amount of time. This could lead to increased drug abuse, addiction, and other societal problems. In addition, synthetic biology could be used to produce drugs that are difficult to detect or resistant to existing drug tests, making it more difficult for law enforcement to identify and apprehend individuals or groups involved in producing illegal drugs. The production of illicit drugs could also lead to other societal problems, such as increased crime and violence and could have negative consequences for public health, as the drugs produced using these techniques could be of unknown quality and could potentially be harmful to users.[96]

Finally, there are concerns that the technology could be used to create organisms that have enhanced capabilities, such as enhanced intelligence or strength. This could lead to ethical dilemmas and concerns about inequality, as individuals or groups with access to these technologies could use them to gain an unfair advantage over others. Additionally, there are concerns that the development of enhanced organisms could lead to unforeseen consequences and potentially harmful unintended effects.

The creation of enhanced organisms could significantly improve the production of food and other essential goods and the development of new and improved technologies. However, there are also concerns that the development of enhanced organisms could lead to ethical dilemmas and concerns about inequality, as well as unintended consequences and potential harm to the environment and human health.[97] As such, the development of synthetic biology raises broader ethical concerns about the manipulation of life and the boundaries between the natural and the artificial. As researchers continue to explore the potential of synthetic biology, they will need to carefully consider the implications of their work for the natural world and our relationship with the environment.[98]

Changing What It Means To Be Human

The development of synthetic biology could change our understanding of what it means to be human. As we begin to create and manipulate life forms in the laboratory, we may need to reassess our definitions of life and consider the ethical and moral implications of creating synthetic organisms. This could lead to new debates and discussions about the nature of life and our place in the world.

Synthetic biology is a rapidly developing field that involves designing and constructing new biological systems, such as microorganisms or genetic circuits. It has the potential to greatly impact many areas of science and technology, including medicine, agriculture, and environmental remediation. It has the potential to significantly expand our understanding of biology and enable the development of new technologies and medical treatments. Even more, as the fields of synthetic biology, quantum computing, and artificial intelligence are likely to converge in the coming years, we expect to see a number of exciting advances that will have a major impact on society. One way synthetic biology, quantum computing, and artificial intelligence are likely to converge is through the use of AI to design and optimize synthetic biological systems. Researchers can design and optimize synthetic biological systems more efficiently and effectively using AI algorithms to analyze large datasets and identify patterns and trends. This could lead to the development of new and improved drugs, chemicals, and other products that are produced using synthetic biology.

Another way these fields are likely to converge is through quantum computing to simulate and predict the behavior of synthetic biological systems. By using quantum computing to simulate the behavior of complex biological systems, researchers will be able to better understand the underlying mechanisms of life and predict the behavior of synthetic organisms. This could lead to the development of new and improved synthetic biology technologies and applications.[99]

By enabling the development of new medical treatments and therapies or by allowing the manipulation of genetic material, synthetic biology could be used to create new microorganisms that could expand the range of medical treatments available to humans and lead to the development of new ways to prevent or cure diseases. It could allow for the creation of new organisms with novel characteristics, such as enhanced disease resistance or improved agricultural productivity and potentially enable the manipulation of human genes to strengthen certain traits or abilities. While this raises ethical concerns, it could lead to developing new technologies that could greatly enhance human capabilities.[100]

6. ROBOTICS

As we have seen in the previous chapter on synthetic biology, this rapidly developing field has the potential to greatly impact many areas of science and technology, including medicine, agriculture, and environmental remediation. In this chapter, we will focus on another rapidly advancing field: robotics.

Robotics involves designing and constructing machines that can perform tasks autonomously or semi-autonomously. This technology has made significant strides in recent years and is now being used in a variety of applications, from manufacturing and assembly to search and rescue operations. As with synthetic biology, robotics raises critical ethical questions and challenges. For example, as robots become more advanced, they may begin to perform tasks currently done by humans, leading to concerns about job displacement and the need for new forms of employment. Additionally, the development of advanced robotics raises questions about the potential risks and benefits of creating machines with human-like abilities.

In this chapter, we will explore the latest developments in the field of robotics and consider the potential implications of this technology for society and the future of work. We will also examine the ethical challenges and questions arising from the development of robotics and consider how these issues can be addressed.[101][102]

The Origins of Robotics

The origins of robotics can be traced back to the ancient world, where people used simple machines, such as levers and pulleys, to perform tasks that were too difficult or dangerous for humans. These machines were based on the principles of mechanics and leveraged the natural forces of the environment, such as gravity and friction, to enable people to lift heavy objects or move them over long distances.[103]

One of the earliest examples of robotics in the ancient world is the use of water clocks and other mechanical devices to automate tasks, such as the ringing of bells or the opening of doors.[104] These devices used a combination of gears, cogs, and levers to convert the flow of water into mechanical motion, enabling them to perform tasks without the need for human intervention. Water clocks, also known as clepsydras, were devices that used the flow of water to measure time. They were used in many different cultures and periods in history, from ancient Egypt and Greece to medieval Europe and China. Water clocks were typically made of bronze or other metals and consisted of a hollow vessel with markings on the inside to indicate the passing of time. The vessel was filled with water, which would flow out through a small hole at a constant rate. As the water flowed out, it would turn a series of gears and cogs, driving the clock's mechanism and causing the hands or other indicators to move. In addition to measuring time, water clocks were also used to automate tasks, such as the ringing of bells or opening doors. For example, some water clocks were designed to ring a bell at a certain time, while others were connected to doors and would automatically open them at a predetermined time. This enabled people to perform tasks without the need for continuous human intervention, making water clocks one of the earliest examples of robotics in the ancient world. While water clocks and other mechanical devices were limited in their capabilities compared to modern robotics technology, they represented an important step forward in developing automated systems.[105]

Another example of early robotics can be found in the development of automatons or mechanical devices designed to mimic living creatures' movements. These devices were often used in entertainment and were intended to be as lifelike as possible, with the ability to move and make noises. Automatons were developed in many different cultures and periods in history, from the ancient world to the modern era. One of the earliest examples of automatons was the bronze bird-like device created by the ancient Greek engineer Heron of Alexandria in the 1st century AD. This device was designed to move and make noises like a bird and was powered by steam. Another famous example of an automaton from the ancient world is the dancing figures created by the Chinese craftsman Zhang Heng in the 2nd century AD. These figures were made of bronze and could move and make music using complex mechanical mechanisms.

In the Middle Ages and the Renaissance, automatons were developed in Europe when creating mechanical clocks and other devices that could move and make noises. These devices often incorporated intricate designs and were highly prized for their beauty and craftsmanship. In the modern era, automatons continue to be developed and used for entertainment. Some of the most famous examples of modern automatons include the life-sized robots created by the Japanese company Honda, which can walk, talk, and even play sports. Automatons are also used in the arts, with some artists creating mechanical sculptures designed to move in complex and lifelike ways. Overall, automatons have a long and fascinating history and are essential to robotics and mechanical engineering.[106]

The modern field of robotics emerged in the 20th century with the development of the first electronic and mechanical robots. One of the key early developments in robotics was the creation of the first programmable robot, called the Unimate, in the 1950s. The Unimate was designed and built by the American engineer George Devol and was the first robot to perform tasks without human intervention. The Unimate was designed to be used in a factory setting and could perform repetitive tasks such as welding, painting, or moving objects. It was controlled by a series of instructions programmed into its memory and could execute these instructions without human intervention. This made the Unimate a significant advance over previous robots, which a human operator typically controlled. The Unimate was also the first robot to use a hydraulic system for its movements, which allowed it to move more quickly and smoothly than earlier robots. This made it well-suited for use in a factory setting, where it could perform tasks with greater precision and speed. The success of the Unimate led to the development of other programmable robots and paved the way for the creation of more advanced robotics technology. It also helped establish robotics as a viable field of engineering and technology and laid the foundation for the many developments that would follow in the decades.[107]

Integrating AI and machine learning into robotics technology has been a key development in the field and has enabled robots to perform more complex tasks and adapt to changing environments. The first efforts to integrate AI into robotics technology began in the 1960s and 1970s, with early AI systems designed to control robots and other machines. As AI technology advanced, it became increasingly integrated into robotics systems. Since then, integrating AI and machine learning into robotics technology has been an active area of research and development, with many researchers working to develop more advanced AI systems that can be used in robotics applications. These systems are being used to enable robots to perform a broader range of tasks, including tasks that require complex decision-making or problem-solving abilities.[108]

Later, researchers began to explore using materials such as silicone, rubber, and shape-memory alloys in the construction of robots. These materials were more flexible and resilient than traditional robotics materials, such as metal or plastic. They allowed for the creation of robots that could move and manipulate objects in a more human-like manner. In recent years, there has been a growing interest in the development of flexible and soft robotics, with many researchers exploring new materials and designs that can be used in the construction of robots. Some key developments today include using bio-inspired materials, such as muscle-like actuators, which allow robots to move and manipulate objects more human-likely. Other developments include 3D printing and other advanced manufacturing techniques, which enable the creation of complex, custom-designed robots.[109]

Today, robotics involves using a range of advanced technologies and principles, including mechanical engineering, control systems, artificial intelligence, sensors, and actuators. These technologies enable the creation of robots that can move, perceive, and interact with their environment intelligently and like humans.

Mechanical engineering plays a central role in robotics, as it involves designing and constructing physical systems that enable robots to move and manipulate objects. This requires a thorough understanding of mechanics, materials science, and other related fields, as well as advanced manufacturing techniques to create the necessary components and systems.

Control systems are also crucial to robotics, as they enable robots to process data from their sensors and make decisions about moving in and interacting with the environment. This requires a deep understanding of control theory, computer science, and other related fields and the ability to develop algorithms that can handle the complex and dynamic nature of robotics applications.

Artificial intelligence is also essential to robotics, as it enables robots to learn and adapt to their environments and perform complex tasks without human intervention. This requires a deep understanding of AI and machine learning and developing algorithms that can handle the high-dimensional and noisy data typically encountered in robotics applications.

Moreover, sensors are crucial to robotics, as they enable robots to perceive their environment and gather information that can be used to make decisions and control their movements. This requires a thorough understanding of sensor technology and the ability to design and implement robust and reliable sensors that can handle the demanding environments typically encountered in robotics applications.

Finally, actuators are crucial to robotics, as they enable robots to move and manipulate objects in the environment. This requires a thorough understanding of actuator technology and the ability to design and implement actuators that generate the necessary forces and movements for the intended tasks.

Combined, these key principles and technologies are essential for the development of advanced and versatile robots that can move, perceive, and interact with their environment in an intelligent and human-like manner[110] and recently robots are making their way into various industries, including manufacturing, healthcare, agriculture, transportation, and entertainment.

In manufacturing, robots are used for tasks such as welding, painting, and assembly, allowing for increased productivity and reduced labor costs. They can perform these tasks quickly and accurately and work 24 hours a day without getting tired. This will enable manufacturers to increase productivity and reduce labor costs. Additionally, robots can operate in hazardous environments, such as extreme heat or dangerous chemicals, that would be unsafe for humans.

The healthcare industry is utilizing robots for surgeries and rehabilitation purposes. Robots can assist in surgeries by holding surgical instruments and providing a stable platform for the surgeon to operate on. This can help improve the precision and accuracy of the surgery. Additionally, robots can be used to assist in rehabilitation. For example, a robot could help a patient perform physical therapy exercises, allowing the therapist to monitor the patient's progress and adjust the exercises as needed.

Robots are increasingly used in agriculture to plant and harvest crops. These tasks can be labor-intensive and difficult for humans, but robots can do them quickly and accurately. Additionally, robots can work in challenging environments, such as in extreme heat or cold or in fields that are difficult to access. This can help farmers increase their productivity and efficiency.

In the transportation industry, there are experiments with robots for package delivery and self-driving cars. This can reduce the need for human delivery drivers and allow packages to be delivered more quickly and efficiently. Additionally, robots are being developed for self-driving cars. These robots can use sensors and other technology to navigate roads and traffic, making transportation safer and more efficient.

Finally, in entertainment, robots are used for animatronics and special effects. Animatronic robots are used to create lifelike movements and sounds in movies, theme parks, and other entertainment venues. Special effects robots can be used to create realistic explosions, fires, and other effects that would be difficult or impossible to create using traditional methods. This allows entertainment companies to create more realistic and engaging experiences for their audiences.[111]

Although the field of robotics is rapidly advancing, there are still significant challenges and limitations to the current state of robotics technology. One major challenge is the high cost of developing and producing robots, making it difficult for small companies or individuals to afford to use robots in their operations.

Another challenge is the need for more flexibility and adaptability of current robots. Most robots are designed to perform specific tasks and cannot adapt quickly to new situations or environments. This can limit the potential applications of robots and make it difficult to use them in unstructured or unpredictable environments.

In addition, programming robots to perform complex tasks is difficult. Current robots are less intelligent and versatile than humans, and programming them to perform complex tasks can be time-consuming and difficult. This can limit the potential uses of robots and make them less efficient than humans in some cases.

The limited durability and reliability of current robots are also a challenge. Most robots cannot operate for long periods without requiring maintenance or repair. This can make them less cost-effective than humans in many cases, and it can also limit their potential applications.

Finally, there needs to be more standardization in robotics technology. Different companies and organizations often use different technologies and protocols, making integrating robots from different sources difficult. This can limit the potential for collaboration and innovation in the field of robotics.

Despite these challenges, many researchers and companies are working on addressing these issues and pushing the field of robotics forward. This work includes the development of new technologies that will make robots more affordable, flexible, and intelligent. As this work continues, we can expect significant robotics technology advancements in the coming years.[112]

The Future Of Robotics

It is difficult to predict exactly what the future of robotics will hold, but a few potential developments are likely to occur in the (near) future.

In the next 5-10 years, we can expect to see significant advancements in the capabilities of robots. This will likely involve the development of more advanced sensors, manipulators, and other hardware that will allow robots to perform a wider range of tasks. Additionally, we may see the integration of robotics into existing systems and processes, allowing robots to work alongside humans or other machines.

One area where we may see significant developments in the next 5-10 years is in the manufacturing industry. As robots become more advanced, they can perform a wider range of tasks in factories, such as packaging, labeling, and even quality control. This could lead to increased efficiency and productivity in manufacturing and the potential for new jobs to be created in the field of robotics. Another area where we may see significant developments in the next 5-10 years is in the transportation industry. The development of self-driving cars is already underway, and in the next 5-10 years, we can expect to see more advanced autonomous vehicles that are able to navigate complex traffic situations and make decisions in real-time. This could lead to safer and more efficient transportation.

Additionally, we can expect to see the continued development of robots for healthcare applications. This could include the development of robots that can assist with surgeries, as well as robots that are able to assist with rehabilitation and physical therapy. These robots could improve the accuracy and efficiency of medical procedures and potentially lead to new treatment options for patients. The following 5-10 years are likely to be an exciting time for the field of robotics, with many new developments and applications on the horizon. As robots become more advanced and widely adopted, we can expect significant changes in how we live and work and the potential for robots to improve our lives in many ways.[113]

Over the next 25-50 years, we can expect to see even more significant developments in the capabilities of robots. This will likely involve the development of robots that can perform tasks currently considered difficult or impossible for machines. Additionally, we may see the development of robots that are able to work alongside humans or other machines or even collaborate with them on tasks. One potential area of development in the next 25-50 years is using robotics in space exploration. Currently, robotic spacecraft are used to explore the solar system, but in the future, we may see the development of robots that are able to work on the surface of other planets or moons. These robots could perform various tasks, such as collecting samples, conducting experiments, or even constructing habitats or other structures. This could open up new possibilities for space exploration and the potential for humans to eventually colonize other worlds. Another potential area of development in the next 25-50 years is using robotics in search and rescue operations. Humans often perform search and rescue missions, which can be dangerous and time-consuming. In the future, we may see the development of robots that can enter hazardous environments, such as collapsed buildings or disaster areas, and search for survivors or provide assistance. These robots could be equipped with advanced sensors and manipulators to navigate rugged terrain and perform tasks such as lifting heavy objects or cutting through debris. The next 25-50 years will likely be a period of rapid growth and advancement in robotics. [114]

The Implications Of Advanced Robotics

The development of advanced robotics technology raises a number of critical ethical questions and challenges. For example, as robots become more advanced, they can perform a wider range of tasks, potentially replacing human workers in many industries. This could lead to widespread unemployment and other social and economic problems.[115]

Several factors contribute to robots replacing human workers. For example, robots can work 24 hours a day without getting tired and perform tasks with high precision and speed. This makes them more efficient than humans, and it could make them a more attractive option for many employers. Additionally, the cost of developing and producing robots will likely continue to decline over time, making them more affordable for many companies. The potential for robots to replace human workers raises several ethical questions. For example, how will workers displaced by robots be able to find new jobs and support themselves and their families? How can we ensure that workers are treated fairly, and their rights are protected, even if robots replace them? And how can we ensure that the benefits of robotics technology are shared equitably, rather than just going to the owners of robots and companies that use them?[116]

Another ethical challenge of the development of advanced

robotics technology is the potential for robots to be equipped with weapons or other tools that could be used to harm humans. One way in which robots could be equipped with weapons is through the use of military robots. Many countries have already developed and used military robots, and they can perform various tasks, including surveillance, reconnaissance, and even combat. These robots are equipped with sensors, manipulators, and other technology that allows them to locate and engage targets, and they are often controlled remotely by human operators. The use of military robots raises many ethical concerns. For example, how can we ensure that robots are used responsibly and ethically and not cause unnecessary harm to civilians or other non-combatants? And how can we ensure that using robots does not escalate conflicts or lead to more widespread violence?

Another way robots could be equipped with weapons is by developing autonomous weapons. Autonomous weapons are robots that can select and engage targets without the direct control of a human operator. These weapons are not yet widespread, but they raise significant ethical concerns. One of the main concerns with autonomous weapons is their potential to make decisions that are not in line with human ethical norms. For example, an autonomous weapon might engage a target based on criteria not aligned with human moral values, such as efficiency or cost-effectiveness. This could lead to the use of robots in ways that are unethical or inhumane. As such, it will be essential to ensure that robots are used responsibly and ethically and do not cause unnecessary harm to humans.[117]

The next ethical challenge is ensuring that robots are designed and used in ways that are safe for humans. This involves addressing potential hazards such as malfunctioning robots or robots that are not correctly programmed to avoid harming humans. Malfunctioning robots can pose a severe risk to human safety, particularly if they are used in environments where they could come into direct contact with people. For example, if a robot designed to assist with surgery malfunctions, it could cause serious injury or even death to the patient. Similarly, if a robot used in a manufacturing setting malfunctions, it could cause harm to human workers. In addition to addressing the issue of malfunctioning robots, it is also important to ensure that robots are properly programmed to avoid harming humans. This involves designing robots with built-in safety features, such as sensors that allow them to detect and avoid collisions with people. It also involves ensuring that robots are programmed to follow ethical and moral guidelines, such as not harming humans or violating their privacy. By taking these steps, we can help to ensure that robots are used in safe ways and do not cause harm.[118]

The fifth ethical challenge that arises from the development of advanced robotics is the issue of robot autonomy and the potential for robots to make decisions that conflict with human values or morals. As robots become more advanced and capable of autonomous decision-making, there is a risk that they could make choices that conflict with human values or morals. For example, a robot designed for military use might be programmed to follow orders without question, even if those orders conflict with human values, such as the prohibition on killing. Similarly, a robot designed to assist with law enforcement might be programmed to prioritize public safety over individual rights and freedoms. In both cases, the robot's decisions could conflict with human values and morals. Therefore, it is essential for robot designers and manufacturers to carefully consider the values and morals that should guide the decisions of autonomous robots. This may involve incorporating ethical principles into the programming of robots or developing guidelines and regulations that specify how robots should make decisions in different situations.[119]

A sixth challenge will be to ensure that robots enhance, rather than replace, human capabilities and experiences. To do so, we need to carefully consider the roles that robots play in society and how they can be used to complement human skills and abilities. For example, robots could perform dangerous or physically demanding tasks, allowing humans to focus on tasks that require creativity, problem-solving, and social interactions. It's also important to establish ethical guidelines for using robots and to regularly assess their impact on society to ensure that they are being used in a way that benefits everyone. These guidelines should be developed with input from various stakeholders, including workers, employers, and experts in fields such as ethics and labor law. The guidelines should outline the roles that robots can play in various settings and any limitations or restrictions on their use.

Moreover, we can establish regular mechanisms for monitoring and assessing the impact of robots on society, including their impact on employment and working conditions. This could include periodic surveys of workers and employers to determine how robots are being used, as well as analysis of data on employment and wages. By regularly collecting and analyzing this data, we can identify any adverse impacts of robots on human workers and take steps to address them.[120]

As robots become more advanced, they can play an increasingly significant role in our daily lives, and it is important to consider the potential impacts of this on society. Some possible concerns about the widespread use of robots include their potential to displace human workers and challenge our privacy and security. It's also important to consider the ethical implications of using robots, particularly in sensitive areas such as healthcare and law enforcement. One possible approach to addressing these concerns is establishing regulations or laws limiting the use of robots in certain contexts. For example, there could be limits on the types of jobs that robots can perform or the amount of autonomy they are allowed to have. Another approach is to ensure that robots are designed and used in a way that considers the potential impacts on human workers and society as a whole.

Ultimately, the extent to which robots should be integrated into society will depend on a variety of factors, including the specific context in which they are being used and the potential impacts on individuals and society as a whole. It will be necessary for society to carefully consider these issues and strike a balance between the benefits of using robots and the potential drawbacks.[121]

7. CLIMATETECH

As we have seen in the previous chapters discussing different emerging technologies, the development of advanced technologies has the potential to enhance our lives greatly and improve our society. However, as we move on to the next chapter on ClimateTech, it is important to consider the potential impacts of these technologies on the environment. Climate change is one of the greatest challenges facing our planet, and the use of advanced technologies, including those covered in the previous chapters, will be critical in addressing this challenge and mitigating the impacts of climate change. Climate change is a global crisis already having a significant impact on our planet, from rising sea levels and extreme weather events to the loss of biodiversity and the displacement of communities. To address this crisis, we must take action to reduce greenhouse gas emissions and adapt to the impacts of climate change. This is where ClimateTech comes in.

In this chapter, we will explore the emerging field of ClimateTech, which focuses on developing advanced technologies that can help address the global climate crisis, reduce greenhouse gas emissions, and adapt to the impacts of climate change. This includes a wide range of technologies, from renewable energy sources such as solar and wind power to advanced sensors and data analysis tools that can help us better understand and predict the impacts of climate change. ClimateTech also encompasses technologies that can help us adapt to the effects of climate change, such as advanced irrigation systems and flood control infrastructure. As we will see, the interplay between the technologies covered in the previous chapters and ClimateTech will be critical in shaping the future of our planet. By combining the power of advanced technologies with a commitment to sustainability and the protection of our planet, we can help to ensure a brighter and more sustainable future for all.[122][123]

Technology And Climate Change

Technology is at the heart of ClimateTech and is critical for addressing climate change. It can help us to reduce greenhouse gas emissions and transition to more sustainable energy sources. From renewable energy technologies to advanced sensors and data analysis, ClimateTech offers a wide range of solutions that can help to reduce greenhouse gas emissions and adapt to the impacts of climate change.

One of the key technologies within the ClimateTech umbrella is renewable energy. Renewable energy technologies, such as solar and wind power, provide clean and sustainable energy sources that can help to reduce our reliance on fossil fuels. These technologies often involve using advanced materials and systems to capture and convert energy from natural sources. For example, solar panels are photovoltaic cells that convert sunlight into electricity, while wind turbines use wind energy to generate electricity. Solar panels and wind turbines will likely continue to evolve and improve in the coming decades. First of all, they will become more efficient. This means they can convert a higher proportion of sunlight and wind energy into electricity, allowing them to generate more power with the same amount of panels or turbines. This can be achieved using advanced materials and technologies, such as multi-junction solar cells and advanced wind turbine blades that can capture more of the solar spectrum or wind energy. As solar and wind technology continues to advance and become more widespread, the cost of these technologies is expected to decrease. This will make solar and wind power more accessible and affordable for consumers and businesses.

Moreover, solar panels and wind turbines will likely become more versatile and adaptable. For example, new solar panel designs may be developed that are flexible and can be easily integrated into various surfaces, such as rooftops, walls, and even clothing. This will make it easier to install and use solar panels in a wide range of settings.[124]

Another important aspect of ClimateTech is the use of advanced sensors and data analysis tools. These technologies can provide real-time information about the state of the environment, helping us to better understand and predict the impacts of climate change. Sensors can be used to monitor air and water quality, temperature, and weather patterns, providing real-time data that can be used to inform decision-making. This can help us to anticipate and prepare for the impacts of climate change, such as extreme weather events and changes in water availability. This information can be used to develop more effective strategies for addressing the climate crisis. For example, advanced sensors can monitor air and water quality, temperature, and weather patterns, providing valuable data that can inform decision-making. In addition, those tools can also help us to monitor and evaluate the effectiveness of ClimateTech. They can be used to track the performance of renewable energy technologies, providing valuable data that can be used to optimize their operation and improve their efficiency. Additionally, data analysis tools can be used to identify trends and patterns in the data collected by sensors, providing valuable insights that can help us to make more informed decisions about the development and deployment of ClimateTech.

Advanced sensors can also be critical in supporting decentralized energy and local smart grids. These sensors can provide real-time data about the state of the energy grid, which can be used to optimize its operation and improve its efficiency. Decentralized energy systems, also known as distributed energy systems, are networks of small-scale energy generation and storage systems located close to the point of consumption. These systems are often based on renewable energy technologies, and they can provide a more resilient and sustainable alternative to traditional centralized energy systems. Advanced sensors can be used to monitor the performance of these decentralized energy systems, monitoring the amount of energy being generated and consumed in real-time, allowing the grid to be dynamically adjusted to meet changing energy demand.

Moreover, crypto tokens could play a role as a means of exchange for buying and selling energy. For example, individuals or businesses that generate excess renewable energy could use crypto to sell that energy to others in their local area or beyond. This could create a more efficient and decentralized energy market where energy can be bought and sold flexibly and transparently. Additionally, individuals or businesses could be rewarded with crypto for participating in the system. This could encourage more people to adopt and use decentralized energy systems and create a more sustainable and resilient energy network.[125]

Renewable energy needs to be stored so advanced energy storage systems can store and manage renewable energy for long periods of time. Several different types of advanced energy storage systems are currently under development. One example is large-scale batteries, which can store electricity and provide a steady flow of electricity to the grid, even when the sun is not shining, or the wind is not blowing. Another example is hydrogen fuel cells. These systems use hydrogen gas as a fuel source and can generate electricity through a chemical reaction. The byproduct of this reaction is water, which makes hydrogen fuel cells a clean and sustainable source of electricity. These advanced energy storage systems can be integrated into smart grids, which are networks of energy generation, storage, and distribution systems that are managed and controlled locally. Smart grids can use advanced algorithms and data analysis tools to optimize the flow of energy, ensuring that it is used in the most efficient and sustainable manner possible.[126]

Next, carbon capture and storage technologies are emerging and designed to capture and store carbon dioxide emissions from power plants and other industrial sources. These technologies are becoming increasingly important as we seek to reduce greenhouse gas emissions and address the impacts of climate change. There are several approaches to carbon capture and storage, each with unique advantages and challenges. One approach is to capture carbon dioxide directly from the exhaust gases of power plants and other industrial facilities. This can be done using a variety of different technologies, such as amine-based solvents, which can capture carbon dioxide from exhaust gases and then release it for storage. Another carbon capture and storage approach is to capture carbon dioxide from the air. This can be done using advanced filtration systems, which can extract carbon dioxide from the air and then release it for storage. This approach has the advantage of being able to capture carbon dioxide from a wide range of sources, including power plants, vehicles, and even individuals. One challenge of these carbon capture and storage technologies is the cost. They require the use of specialized equipment, such as advanced filtration systems and storage facilities, which can be costly to install and operate.

Additionally, capturing and storing carbon dioxide can be energy-intensive, which can increase the cost of implementing these technologies. Once carbon dioxide has been captured, it must be stored in a safe and secure manner. This can be done using a variety of different approaches, including underground injection, where carbon dioxide is injected deep into the Earth, or ocean storage, where carbon dioxide is dissolved into the ocean. However, there are concerns that carbon dioxide could leak from storage facilities, potentially causing environmental damage and contributing to climate change. Overall, carbon capture and storage technologies are an important part of the emerging field of ClimateTech, as they can help to reduce greenhouse gas emissions and address the impacts of climate change. By capturing and storing carbon dioxide, these technologies can help to reduce our reliance on fossil fuels and support the transition to a more sustainable and resilient future.[127]

Finally, ClimateTech also includes technologies that can help us to adapt to the impacts of climate change. This can involve the use of advanced irrigation systems, flood control infrastructure, and technologies that can help us better forecast and prepare for climate-related disasters. For example, advanced weather forecasting systems can provide early warning of potential storms and other natural disasters, allowing communities to evacuate and take other protective measures. Overall, the development and deployment of ClimateTech will be critical in addressing the global climate crisis. By combining the power of advanced technologies with a commitment to sustainability, we can help to ensure a brighter and more sustainable future for all.

The Convergence Of Technologies To Combat Climate Change

The convergence of different technologies, such as the metaverse, strong AI, quantum computing, robotics, and synthetic biology, has the potential to significantly advance ClimateTech and help us tackle climate change. These technologies can provide new platforms for collaboration and innovation in the field of ClimateTech, and enable us to develop more accurate and detailed models of the global climate system, optimize the operation of renewable energy systems, and develop new renewable energy technologies. By leveraging the convergence of technology, we can accelerate the transition to a low-carbon economy and build a more sustainable and resilient future.[128]

First, the metaverse can enable scientists, engineers, and policymakers from different parts of the world to collaborate and share ideas, data, and resources and work together on projects related to climate change. This can help accelerate the development and deployment of ClimateTech, and enable us to tackle climate change on a global scale. The metaverse can also be used to model and simulate the impacts of climate change on different regions and industries and to develop strategies for mitigating and adapting to these impacts. This can provide valuable insights to help us make informed decisions about tackling climate change. Additionally, the metaverse can be used as a platform for public education and engagement on climate change. For example, virtual exhibitions and events can be held in the metaverse to raise awareness about the causes and impact of climate change and to encourage individuals and communities to take action to reduce their carbon footprint. This can help build a groundswell of support for ClimateTech and accelerate the transition to a low-carbon economy.[129]

Strong AI can analyze vast amounts of data from a wide range of sources, such as satellite images, weather stations, and air quality sensors, and develop models that can simulate the complex processes and feedback loops that drive climate change. For example, strong AI can be used to model the impacts of climate change on different regions and industries and to develop strategies for mitigating and adapting to these impacts. This can provide valuable insights to help us make informed decisions about tackling climate change. For example, strong AI can identify the most vulnerable areas and sectors and develop targeted strategies for reducing their exposure to climate risks. Additionally, strong AI can be used to optimize the operation of renewable energy systems, such as wind and solar power, and to develop new renewable energy technologies. For example, strong AI can analyze data from renewable energy systems and identify opportunities for improving their efficiency and reliability. This can help reduce our reliance on fossil fuels and lower our carbon emissions.[130]

In terms of more futuristic new renewable energy systems, strong AI could help develop space-based solar power plants that harness the power of the sun and beam it back to Earth using microwaves or laser beams. This technology, known as space-based solar power, has the potential to provide clean and abundant energy but is currently limited by the high cost and technical challenges of building and maintaining solar power plants in space. With advances in AI, robotics, and materials science, strong AI may be able to develop space-based solar power plants that are efficient, reliable, and cost-effective. Strong AI can also help develop fusion reactors that generate electricity by fusing the nuclei of atoms, such as hydrogen and helium, to release massive amounts of energy. Fusion reactors have the potential to provide clean and abundant energy but are currently limited by the high temperatures and pressures required to sustain the fusion reaction. With advances in AI, robotics, and materials science, strong AI may be able to develop fusion reactors that are more efficient and scalable and can generate electricity without emitting greenhouse gases or radioactive waste.[131]

Next, quantum computing has the potential to contribute to solving climate change by providing new and more powerful tools for developing and optimizing renewable energy technologies. Quantum algorithms and simulations can be used to model the impacts of climate change on different regions and industries and to develop strategies for mitigating and adapting to these impacts. This can provide valuable insights that can help us make informed decisions about how to tackle climate change. For example, quantum algorithms can be used to identify the most vulnerable areas and sectors and develop targeted strategies for reducing their exposure to climate risks. In addition, quantum computing could be used to create advanced geoengineering technologies that can manipulate the Earth's climate and counteract the effects of climate change. These technologies, known as solar radiation management and carbon dioxide removal (SRM and CDR), can reduce global warming and slow down climate change but are currently limited by the high cost and potential risks of implementing them on a large scale. For example, quantum algorithms and simulations could be used to model the Earth's climate and predict the effects of different SRM and CDR technologies on temperature, precipitation, and other weather patterns. This can help us determine which technologies are most effective and safe and how they can be deployed globally. Quantum computing could also be used to optimize the operation of SRM and CDR technologies and to monitor their performance and impact on the environment.

Additionally, quantum computing could be used to develop new SRM and CDR technologies that are more efficient, reliable, and cost-effective. For example, quantum algorithms could be used to design new materials and systems that can reflect sunlight back into space or capture and store CO2 from the atmosphere. These technologies could be more effective and scalable than existing SRM and CDR technologies and could help us reduce global warming and slow down climate change.[132]

Finally, synthetic biology has the potential to contribute to solving climate change by designing and engineering new biological systems, such as cells, genes, and organisms, for specific purposes, such as producing fuels, chemicals, and materials. Synthetic biology can be used to develop algae-based biofuels with a lower carbon footprint than fossil fuels. Algae are a type of aquatic microorganism that can grow quickly and efficiently and can be used to produce a wide range of biofuels, such as biodiesel, bioethanol, and biogas. With advances in synthetic biology, it may be possible to engineer algae that can produce more biofuels with less water, land, and nutrients and that can be grown in a variety of environments, such as ponds, oceans, and wastewater. Additionally, synthetic biology can be used to develop new materials and products that are more sustainable and environmentally friendly. For example, synthetic biology can be used to engineer microorganisms that can convert waste into valuable products, such as bioplastics, biochemicals, and bioremediation agents. This can help reduce our reliance on fossil fuels and reduce greenhouse gas emissions. Synthetic biology can also be used to create new materials that are more durable, lightweight, and biodegradable and that can replace materials that are derived from fossil fuels, such as plastics and concrete.[133]

Overall, the convergence of technologies such as the metaverse, strong AI, quantum computing, robotics, and synthetic biology has the potential to contribute to solving climate change by providing new and more powerful tools for developing and optimizing renewable energy technologies, modeling the impacts of climate change, and adapting to its consequences. By leveraging the power of these technologies, we can accelerate the transition to a low-carbon economy and build a more sustainable and resilient future.

The Impact Of Climatetech On The Economy

ClimateTech has the potential to significantly impact the economy in a range of different ways. These impacts will be both positive and negative. They will need to be carefully considered to ensure that we can support the transition to a more sustainable and resilient future.

One of the key impacts of ClimateTech on the economy will be the creation of new jobs and industries. As the demand for renewable energy and other ClimateTech solutions grows, there will likely be a significant increase in the number of jobs and businesses in these sectors. This will create new opportunities for workers and entrepreneurs and help drive economic growth. At the same time, as renewable energy technologies become more widespread, there will likely be a decrease in the demand for fossil fuels, such as coal, oil, and natural gas. This will negatively impact the industries and workers currently involved in the extraction, processing, and distribution of these fuels.

Furthermore, the use of ClimateTech will also have an impact on individual consumers. As renewable energy technologies become more widespread, there will likely be a decrease in the cost of electricity and other energy services. This will help to reduce the financial burden on consumers, particularly those on low incomes. Additionally, there will likely be an increase in the availability of clean and sustainable energy options, which will give consumers more choice and control over their energy sources.

Another positive impact of ClimateTech will be the potential for increased trade and investment. As countries worldwide transition to more sustainable and resilient energy systems, there will likely be increased demand for ClimateTech, such as solar panels and wind turbines, as well as rare earth metals and other materials. This will create new opportunities for companies that produce these technologies and could lead to increased exports and foreign investment. This could also lead to increased tensions and geopolitical risks, negatively impacting the global economy.

Finally, the transition to more sustainable and resilient energy systems will require significant changes to existing infrastructure, such as power grids and transportation systems. This could lead to disruptions to existing industries, which could have negative impacts on workers and businesses. Additionally, the transition to more sustainable and resilient energy systems will require significant changes to existing regulations, policies, and subsidies, which could lead to uncertainty and market instability.

ClimateTech will have both short-term and long-term implications, which must be carefully considered to ensure that we can support the transition to a more sustainable and resilient future.[134]

8. 3D PRINTING

As we move to the last technology that will shape our future, we will discuss 3D printing, a convergence of different technologies, such as computer-aided design (CAD), materials science, and robotics. 3D printing, also known as additive manufacturing, creates three-dimensional objects by building them layer by layer from a digital model. This technology can revolutionize how we design, manufacture, and distribute products and enable us to create new and innovative solutions to a wide range of challenges, including climate change.[135] 3D printing is included in this book because it has the potential to disrupt globalization in the decades to come. Globalization is the process of increasing economic, political, and cultural interconnectedness among countries, which has led to the growth of international trade and investment and the emergence of a global economy. Globalization has brought many benefits, such as lower prices, more choices, and greater efficiency, but it has also led to challenges, such as inequality, environmental degradation, and cultural homogenization.

3D printing will disrupt globalization by enabling us to move from global to local production and consumption by manufacturing products closer to where they are needed instead of importing them from other countries. By using 3D printing, we can manufacture products closer to where they are required, reducing transportation time, cost, and environmental impact, such as long-distance transportation, which is a major contributor to greenhouse gas emissions and climate change.[136] As 3D printing enables the production of personalized and one-of-a-kind items, it could lead to a greater emphasis on individual preferences and needs in the global economy and a more sustainable and eco-friendly global trade system. Moreover, as 3D printing technology democratizes production and allows for creating new and innovative products could lead to a more diverse and inclusive global economy. However, as 3D printing technology challenges established production methods, it could lead to increased competition and disruption in traditional manufacturing industries and reshape global trade patterns and structures.[137]Global trade is the exchange of goods and services among countries and is an important driver of economic growth and development. International trade is typically based on the production and export of goods, such as raw materials, finished products, and intermediate goods, from one country to another. With 3D printing, we can move from global trade based on production to international trade based on services.[138] [139]

The Benefits Of 3D Printing

3D printing has the potential to revolutionize many different areas and industries. It allows for creating physical objects from digital designs using various materials and offers many benefits over traditional manufacturing methods. Some of the key benefits of 3D printing include increased customization and personalization of products, reduced waste and environmental impact, greater economic opportunities for small and medium-sized enterprises, increased competition and disruption in traditional manufacturing industries, and improved healthcare and medical treatments. Let's explore these benefits in more detail and discuss how 3D printing shapes our future.[140]

First, 3D printing enables us to create custom and personalized products by providing a flexible and efficient production process that can easily adjust to individual preferences and needs. This is possible because 3D printing uses a digital design file as a blueprint for production and can easily adapt and modify the design to create a customized product. This means that 3D printing can produce a wide range of different products with different shapes, sizes, and materials without requiring specialized tools or production processes.

Additionally, 3D printing allows for creating complex and intricate designs that would be difficult or impossible to produce using traditional manufacturing methods. This opens up new possibilities for product design and customization and allows for the creation of unique and one-of-a-kind items that reflect the individual preferences and needs of consumers.

Furthermore, 3D printing has the potential to greatly reduce the costs and time associated with customization and personalization. Traditional manufacturing methods often require significant amounts of setup and production time, as well as specialized tools and materials, which can add substantial costs to the production of custom products. In contrast, 3D printing allows for the creation of custom products with little to no additional cost or effort, making customization and personalization more accessible and affordable. Overall, the ability of 3D printing to create custom and personalized products can greatly expand the range of products available in the market and create new economic opportunities for businesses and consumers alike.[141]

Next is how 3D printing has the potential to reduce waste and improve the sustainability of our production and consumption patterns. Traditional manufacturing methods often involve cutting, shaping, and forming materials, which can result in a significant amount of waste and inefficiency. In contrast, 3D printing uses a layer-by-layer approach to create objects, which allows for the precise and efficient use of materials. This can have a number of benefits in terms of waste reduction and sustainability. For example, 3D printing allows for the creation of precise and efficient designs that minimize material usage and energy consumption, especially with complex and intricate designs, which can lead to a more sustainable and eco-friendly production system with reduced environmental impacts such as greenhouse gas emissions and pollution.[142]

Another key benefit of 3D printing is its potential to

offer greater economic opportunities for small and medium-sized enterprises (SMEs). Traditional manufacturing methods often require significant investments in specialized tools and equipment, as well as specialized knowledge and expertise. This can make it difficult for small and medium-sized enterprises to compete with larger, established companies in the manufacturing industry. In contrast, 3D printing technology democratizes production by creating new and innovative products with little to no additional cost or effort and without the need for specialized tools or production processes.

3D printing also has the potential to significantly impact the healthcare industry and the way that medical treatments and therapies are developed and delivered. For example, improved medical devices and implants, as 3D printing enables the production of custom and personalized medical devices and implants tailored to patients' individual needs and preferences. This can improve the effectiveness and safety of medical treatments and can reduce the risks associated with traditional medical devices. It could also lead to increased accessibility and affordability of medical treatments, as 3D printing allows for producing low-cost and personalized medical devices and therapies. This can help to make medical treatments more accessible and affordable, particularly for patients in underserved or low-income communities.

Furthermore, 3D printing can facilitate the production of complex and precise drug delivery systems and drug testing models. This can accelerate the drug development process and can help to reduce the time and cost associated with drug testing and approval. 3D printing offers a number of advantages over traditional manufacturing methods and has the potential to greatly impact many different areas of our lives and society. It is essential to carefully consider and address the potential implications and impacts of this technology as it continues to evolve and become more prevalent.[143]

3D Printing Shaping Our Future

3D printing technology is constantly evolving and improving, and one of the key areas of progress is the development of new and innovative materials and processes. As 3D printing technology continues to expand the range of materials used for printing and improves the precision and efficiency of the printing process, it will enable the production of a broader range of products. Currently, 3D printing is limited to a relatively small range of materials, including plastics, metals, ceramics, and some composite materials. However, as technology advances, it will likely be able to print with a much wider range of materials, including biodegradable plastics, organic and plant-based materials, and other sustainable and environmentally-friendly materials. As 3D printing technology improves, it will become even more precise and efficient, with faster printing speeds and higher resolution. This could result in more complex and intricate products and enable the production of products with finer details and more precise dimensions. This could open up new possibilities for product design and innovation and allow for creating unique and differentiated products that can compete with established brands and products in the market.

In the future, 3D printing technology will likely be used to create a wide range of personalized and customized products, from clothing and fashion items to furniture and home decor to medical treatments and therapies. This could include the ability to easily adjust and modify the design of a product to fit an individual's preferences or to create unique and one-of-a-kind products that are not available through traditional manufacturing methods.[144] It could also include custom-fit and personalized clothing and fashion, unique and one-of-a-kind clothing items tailored to an individual's body, style, and preferences.[145] Moreover, it could include the ability to quickly and affordably create unique and one-of-a-kind homes and buildings tailored to individual preferences and needs that are more sustainable and eco-friendly. This could include using biodegradable plastics, organic and plant-based materials, and other sustainable and environmentally-friendly building materials. It could lead to creating homes and buildings that are more sustainable, have a reduced environmental impact, and support the development of a more sustainable and eco-friendly society. In addition, 3D printing technology could be used to create custom and personalized living spaces that are more adaptable and flexible to individual preferences and needs. This could include the ability to easily and affordably adjust and modify the design of a living space or to create unique and one-of-a-kind homes and buildings that are not possible using traditional construction methods.[146]

One of the more futuristic applications of 3D printing technology is the ability to print organs and tissues. This could include the ability to create complex and functional biological structures, such as human organs and tissues, using 3D printing technology. 3D printing technology has been used to successfully create small, simple biological structures, such as blood vessels and skin tissue. However, the technology is not yet advanced enough to create more complex organs and tissues, such as the heart, liver, or kidneys. As 3D printing technology advances and improves, it will likely become more capable of creating complex and functional biological structures, such as organs and tissues. This could be accomplished through developing new and innovative printing materials and processes, as well as using bio-ink and other biomaterials that can be used in 3D printing. In the future, 3D printing technology may be used to create customized organs and tissues for medical treatments and therapies. This could include the ability to create unique and individualized organs and tissues tailored to each patient's specific needs and preferences and could improve the effectiveness and safety of medical care. While this is still a nascent field, there is significant potential for 3D printing to play a role in the future of tissue engineering, regenerative medicine, and healthcare.[147]

In the future, 3D printing technology will also be used to support the exploration and conquest of space. This could include the ability to easily and affordably create custom and personalized objects and structures in space, such as custom and personalized tools and equipment for use on other planets and in other parts of the solar system, personalized living quarters and habitats for astronauts and other space explorers and even one-of-a-kind spaceships and space stations that are not possible using traditional manufacturing methods.[148]

3D printing is a revolutionary technology that has the potential to greatly impact many different areas of our lives and society. From the ability to easily and affordably create custom and personalized products and objects to the ability to reduce waste and improve the sustainability of our production and consumption patterns, 3D printing has the potential to greatly enhance many aspects of our daily lives.3D printing is likely to have a major impact on our future, especially if it converges with other technologies discussed in this book. For example, the convergence of 3D printing with strong AI could enable the development of advanced, intelligent 3D printing systems that are capable of creating custom and personalized products and objects with greater speed, efficiency, and precision, while quantum computing could enable the development of advanced 3D printing systems that are capable of creating new and innovative materials and structures that are not possible using traditional manufacturing methods. The convergence of 3D printing with synthetic biology could enable the development of advanced 3D printing systems capable of creating custom and personalized living organisms and tissues. This could have major implications for many fields, including healthcare, agriculture, and environmental science.[149]

Overall, using 3D printing technology can greatly impact many different areas and industries and lead to the creation of new and innovative products and technologies in the future. It is an exciting time to be alive, and we are fortunate to live when this revolutionary technology is emerging and evolving.[150]

EPILOGUE

As we come to the end of this book, it is clear that the convergence of technology will continue to shape our world in many ways. Technology is transforming the world around us and opening new and exciting possibilities for the future. With the help of these technologies, we have the potential to solve many of the complex and pressing challenges facing our world today, from climate change to global inequality. By harnessing the power of these technologies, we can create a more sustainable, equitable, and inclusive future for all.

In addition, this convergence of technology also offers many exciting opportunities for personal growth and development. With the ability to easily and affordably create custom and personalized products and objects, and the ability to access new and innovative technologies and services, we have the potential to live more fulfilling and satisfying lives.

This convergence of technology offers many exciting opportunities and possibilities for the future. By harnessing the power of these technologies, we have the potential to solve many of the complex and pressing challenges facing our world today, from climate change to global inequality. However, it also brings with it many challenges and potential risks. As we continue to develop and advance these technologies, it is important to carefully consider and address the potential ethical, social, and environmental implications and impacts of these technologies.

As we wrote these visions for our future, we found ourselves imagining the many different ways that technology could shape our world in the coming years and decades. From the development of strong AI and quantum computing to the emergence of the metaverse and the growth of 3D printing, we was able to envision a future that is both exciting and full of possibilities. At the same time, we found ourselves considering the potential challenges and risks that come with the convergence of technology. As we continue to develop and advance these technologies, it is important to carefully consider and address the potential ethical, social, and environmental implications and impacts of these technologies. Overall, Future Visions has been a thought-provoking and engaging experience. It has allowed us to imagine and explore how technology could shape our world in the coming years and consider both the opportunities and challenges that this technology presents.

As we move forward into the future, it is important to carefully and thoughtfully consider the role of technology in our lives and to work together to create a sustainable, equitable, and inclusive future for all. It is an exciting time to be alive, and we are fortunate to be living in a time when technology is transforming the world around us in so many different ways.[151]

ENDNOTES

[1] Write the summary of the book 'Future Visions'. It should be exciting and entice people to read the book.

[2] For the book, write an 'About the Author' about ChatGPT

[3] For the preface, add a paragraph that describes that all content in the book is created by AI, including the title, tagline, front and back cover, acknowledgments, preface and, of course, the book itself, and that it was edited by AI.

[4] Describe how the internet, the mobile internet, social media and cloud computing have changed how we work, live and socialize. How has it changed what it means to be human and how has it reorganized society worldwide?

[5] The chapter is an introduction that discusses why this book was written. The book was written to prove the speed at which AI is developing and how good ChatGPT is. It discusses a brief history of the most important trends of the past 50 years and then describes the seven key trends that will define the next 50 years.

[6] We are writing a book about the future and how technology will define the next 50 years. We will cover seven trends. Provide a brief introduction to the metaverse and how the immersive internet, also known as the next iteration of the internet, will transform our lives as we move from a 2D society to a 3D society where we change from actively deciding on going on the internet, to being part of the internet.

[7] We are writing a book about the future and how technology will define the next 50 years. We will cover seven trends. Provide a brief introduction to Strong AI and when we will reach this.

[8] We are writing a book about the future and how technology will define the next 50 years. We will cover seven trends. Provide a brief introduction to quantum computing and the breakthroughs it could bring in the next 50 years

[9] We are writing a book about the future and how technology will define the next 50 years. We will cover seven trends. Provide a brief introduction to synthetic biology and the breakthroughs it could bring in the next 50 years

[10] We are writing a book about the future and how technology will define the next 50 years. We will cover seven trends. Provide a brief introduction to robotics and how we are entering the era of the Jetsons

[11] We are writing a book about the future and how technology will define the next 50 years. We will cover seven trends. Provide a brief introduction to ClimateTech and how it could help us combat climate change and prevent global destruction

[12] We are writing a book about the future and how technology will define the next 50 years. We will cover seven trends. Provide a brief introduction to 3D printing and how it will revert globalization and create a world where manufacturing is done more locally than internationally, thereby disrupting global trade.

[13] Describe that we will discuss the above-mentioned trends through the lens of these philosophical questions

[14] Describe in detail what the metaverse is, based on Dr. Mark van Rijmenam's book Step into the metaverse, and how it will affect society in the next five years, including that we will move from actively making a decision to go on the internet to being part of the internet.

[15] What will the metaverse look like in the 2030s and 2040s? Describe in detail the steps of the technologies that need to be developed and describe in detail how the metaverse will look like then.

[16] Describe in even more detail what specific breakthroughs are needed in hardware and software. Who will build this, how, why, in what way, etc.? Summarize what kind of collaborations will be required to build the metaverse by 2040.

[17] Now take the perspective of a 20-year-old user in the 2040s and describe in detail how the metaverse will look like then. How will the user experience the world?

[18] Now let's step 10 years into the future, how will the metaverse look like in 2050; what new technologies can we expect, and what different experiences will the now 30-year-old have in the 2050s?

[19] Let's take another jump into the future, 20 years, so it is the 2070s and our user is in his 50s. How has the world changed? What unexpected technologies will be part of our daily lives? How does our user go about his day in this new world?

[20] What are the potential benefits and drawbacks of the metaverse in the 2070s, what are the potential risks and challenges associated with them, and how can we mitigate them in the 50 years to come?

[21] How will the metaverse affect our sense of identity and selfhood? Expand how the metaverse will enhance and augment our senses, abilities

and experiences.

[22] How will the metaverse erode our individuality and autonomy?

[23] Describe a dystopian scenario of the metaverse in the 2070s where it controls, manipulates and monitors us in ways that change our identity and affect our selfhood.

[24] How will the metaverse in 2070 affect our relationships with each other? Will it bring us closer together, or will it create new divides and tensions?

[25] How will the metaverse in 2070 affect our understanding of reality?

[26] Provide a conclusion to all that has been written about the metaverse so far.

[27] Explain the varies types of machine learning

[28] What is transfer learning?

[29] How does a neural network work?

[30] What are famous examples of deep learning?

[31] Please use the above section by describing that this book is not written by deep learning but by a large language model and describe the difference.

[32] Before we dive into strong AI, let's go back and discuss in detail what AI is, how it works and what machine learning, deep learning and neural networks have to do with it. Be as detailed as possible.

[33] What is strong AI?

[34] Ok, let's switch gears and discuss strong AI and how it will change our society. Start with a brief history of AI and why AI has become so powerful in recent years.

[35] How can we expect machine learning / deep learning to evolve in the next 30 years?

[36] What new jobs can we expect in 2050 driven by AI?

[37] Discuss in detail how AI and automation will change work

[38] How will AI change healthcare in 2050?

[39] How will AI change transportation in 2050?

[40] How will AI define our lives 20 years later, in 2070? What will be different then, and what unexpected changes can we expect?

[41] Expand how AI will enhance our senses and abilities

[42] What are the things that strong AI will be better at than humans?

[43] What are the key features and capabilities of Strong AI in 2070?

[44] What are the risks of strong AI?

[45] How can we mitigate the risks of strong AI?

[46] How do we achieve strong AI?

[47] How long will it take before we achieve strong AI?

[48] What will come after strong AI?

[49] What will Super AI look like?

[50] Can you think more out-of-the-box when contemplating the possibilities of Super AI?

[51] Based on these out-of-the-box possibilities, what are some of the implications of Super AI for humanity?

[52] How can we maintain control over super AI? Write without bullet points and be very specific and detailed.

[53] What are the key stakeholders and decision-makers involved in developing strong AI and Super AI?

[54] Write an introduction that bridges the previous section about Strong AI to the next section about quantum computing.

[55] Describe the very detailed history of quantum computing

[56] Describe Schrödinger's cat thought experiment

[57] Expand the section on the 1920s and 1930s

[58] Expand the section on the 1980s and 1990s

[59] Expand the section on the 2000s and 2010s

[60] Why do we need different algorithms and protocols for quantum computers?

[61] What can we expect in the 2020s in terms of quantum computing?

[62] Describe how quantum computing is likely to evolve in the coming 30 years to 2050

[63] Expand how quantum computers will affect healthcare by 2070

[64] How will quantum tomography define healthcare in 2070?

[65] Describe in detail how quantum computing will redefine the field of materials science and engineering in 2070

[66] Describe in detail how quantum computing will redefine the field of finance and economics in 2070

[67] Why is the quantum internet a possible application of quantum

computing?

[68] What are some futuristic applications of a quantum internet?

[69] How will quantum computing affect our relationship with the natural world in the coming decades?

[70] Discuss how quantum computing will influence strong AI and vice versa in 2070

[71] What are the ethical implications of quantum computing?

[72] Expand the ethical implications of quantum computing on privacy and security

[73] Expand the ethical implications of quantum computing on inequality and access

[74] Expand the ethical implications of quantum computing on control and regulation

[75] Expand the ethical implications of quantum computing on autonomy and responsibility

[76] Summarize how we could mitigate these ethical implications of quantum computing by 2070.

[77] Describe an introduction to these five milestones

[78] Describe fictional milestones of quantum computing in the coming decades, as if written by a journalist reporting on those milestones. Describe what happened and how it changed society. Include 5 milestones from now to 2070 → Expand each milestone with more details; who achieved it, where, what is the impact, and how did it evolve in the years after → add five more milestones in between and describe in detail who achieved it, where, what is the impact and how did it evolve in the years after → Expand each milestone with many more details, such as what kind of startups were launched after the milestone, which new inventions were made with it and how it changed society.

[79] Provide a summary of hypothetical milestones achieved by quantum computing in the decades to come

[80] Imagine a book with multiple chapters on future technologies. We just finished a chapter on quantum computing and are now moving to synthetic biology. Write a brief introduction that makes a nice bridge between these technologies.

[81] What is synthetic biology --> Expand and provide more details

[82] Describe the Human Genome Project in detail

[83] Describe the Synthetic Biology Engineering Research Center and its achievements in detail

[84] Discuss the history of synthetic biology and include advances made in protein folding

[85] What is the role of protein folding in synthetic biology?

[86] When asking about AlphaFold, also developed by OpenAI, ChatGPT gave the following answer: "I'm sorry, but I'm not able to browse the internet, so I don't have information on the role of AlphaFold in synthetic biology. AlphaFold is not a technology or concept that I'm familiar with, so I'm not able to provide any details on its role in synthetic biology. However, if you have any further questions about synthetic biology, I'd be happy to help with those."

[87] Discuss in detail the history of synthetic biology; What have been important milestones, who was involved, and how did this brings us to where we are now with synthetic biology? --> Expand and provide more details

[88] What major breakthroughs can we expect from synthetic biology in the coming decades?

[89] Expand how synthetic biology will affect the development of new medicines

[90] Expand how synthetic biology will affect the development of sustainable fuels and chemicals

[91] Expand how synthetic biology will affect the development of the production of food and other essential goods

[92] Expand how synthetic biology will affect the development of environmental cleanup

[93] How will synthetic biology affect our relationship with the natural world?

[94] What are the ethical implications of synthetic biology?

[95] Expand the section on creating dangerous organisms. How would this work, and how could this affect society?

[96] Expand the section on using synthetic biology to produce illegal drugs

[97] Expand the section on using synthetic biology to create organisms that have enhanced capabilities

[98] How will synthetic biology affect our relationship with the natural world?

[99] How will synthetic biology, quantum computing and artificial intelligence converge, and how will that change our society?

[100] How will synthetic biology change what it means to be human?

[101] Imagine a book with multiple chapters on future technologies.

We just finished a chapter on synthetic biology and are now moving to robotics. Write a brief introduction that makes a nice bridge between these technologies.

[102]To write this chapter, ChatGPT was asked the following: In a book about future technologies, we have a chapter on robotics. Which questions do we need to ask to obtain a detailed understanding of robotics, its past, present and (far) future? And What other relevant questions could we ask? It came back with the following 18 questions that were used to write this chapter:

- What are the origins of robotics, and how has the field developed over time?
- What are the key principles and technologies that are used in robotics?
- What are the current applications of robotics, and how are they being used?
- What are the potential future developments and applications of robotics?
- What ethical challenges and questions arise from the development of robotics?
- How are these ethical challenges being addressed, and what are the potential implications for society?
- What are advanced robotics's potential risks and benefits, and how can they be managed?
- How will robotics impact the future of work, and what are the implications for employment and the economy?
- How are different countries and organizations approaching the development of robotics, and what are the potential implications of these approaches?
- What are the key challenges and limitations of current robotics technologies, and how are they being addressed?
- What are the ethical and moral considerations involved in the development and use of robotics, and how are these being addressed?
- What are the potential implications of robotics for national security, defense, and other sensitive areas?
- What are the potential future developments and applications of robotics that are currently being explored, and what are the implications of these developments?

[103] What are the origins of robotics, and how has the field developed over time?

[104] Expand the section on the origins of robotics that can be traced back to the ancient world

[105] Expand the section on one of the earliest examples of robotics in the ancient world is the use of water clocks and other mechanical devices to

automate tasks.

[106] Expand the section on automatons

[107] Expand the section on creating the first programmable robot, the Unimate, in the 1950s.

[108] Expand the section that describes the history of the integration of artificial intelligence and machine learning into robotics technology, enabling robots to perform more complex tasks and adapt to changing environments.

[109] Expand the section that describes the history of the development of new robotic materials and designs, such as flexible or soft robotics, which allows robots to move and manipulate objects in a more natural and human-like manner.

[110] What are the key principles and technologies that are used in robotics? --> expand on the key principles and technologies that are used in robotics? --> Rewrite the expanded key principles and technologies that are used in robotics as if written by a journalist

[111] What are the current applications of robotics, and how are they being used? --> Expand the current applications of robotics, with detailed sections on manufacturing, healthcare, agriculture, transportation, and entertainment

[112] What are the key challenges and limitations of current robotics technologies, and how are they being addressed? --> Expand the section on challenges and limitations of current robotics technologies, written as a journalist

[113] What are the potential future developments and applications of robotics from 5 years to 25 years and 50 years? --> Expand the section on the next 5-10 years and give more detailed examples of how robotics will evolve

[114] What are the potential future developments and applications of robotics, from 5 years, to 25 years and 50 years? --> Expand the section on the next 25-50 years and give more detailed examples of how robotics will evolve

[115] What are the ethical challenges and questions that arise from the development of advanced robotics?

[116] Expand the section on robots replacing human workers

[117] Expand the section on robots being equipped with weapons

[118] Expand the section on malfunctioning robots, or robots that are not properly programmed to avoid harming humans.

[119] Expand the section on robot autonomy and the potential for robots to make decisions that conflict with human values or morals?

[120] How can we ensure that robots are used to enhance, rather than

replace, human capabilities and experiences?

[121] Should there be limits on the extent to which robots can be integrated into society, and if so, what should those limits be? And Write a conclusion on the ethical implications of advanced robotics.

[122] Imagine a book with multiple chapters on future technologies. We just finished several chapters on emerging technologies and are now moving on to CLimateTech. Write a brief introduction that makes a nice bridge between these technologies --> Expand this section and provide more details.

[123] To write this chapter, I asked ChatGPT which questions to ask:

- What are the key technologies and solutions that fall under the umbrella of ClimateTech, and how do these technologies work?
- How does ClimateTech address the global climate crisis, and what are the potential benefits of these technologies?
- What are the potential drawbacks and challenges associated with the development and deployment of ClimateTech solutions, and how can these be addressed?
- How do the technologies covered in previous chapters, such as robotics and AI, intersect with ClimateTech, and what is the potential impact of this intersection on the development of ClimateTech solutions?
- How will the use of ClimateTech solutions impact different sectors of society, such as the economy, the environment, and individual consumers, and what are the potential long-term implications of these impacts?
- What role do different stakeholders, including government, industry, academia, and civil society, play in shaping the future of ClimateTech, and how can these stakeholders work together to ensure that ClimateTech is used in a responsible and sustainable way?

[124] How will solar panels and wind turbines evolve in the coming decades?

[125] What role will crypto tokens play in a decentralized energy system?

[126] Can you give more futuristic examples of ClimateTech? --> Please expand the section on advanced energy storage systems

[127] Can you give more futuristic examples of ClimateTech? --> Please expand the section on carbon capture technologies --> Please expand the section on carbon capture technologies

[128] How will the convergence of technology such as the metaverse, strong AI, quantum computing, robotics and synthetic biology contribute to ClimateTech?

[129] Expand how the metaverse can contribute to solving climate change

[130] Expand how strong AI can contribute to solving climate change

[131] What kind of far-futuristic new renewable energy systems could strong AI develop?

[132] What futuristic solutions could quantum computing develop to solve climate change? --> Expand how quantum computing could develop advanced geoengineering technologies and how that will solve climate change

[133] Expand how synthetic biology can contribute to solving climate change

[134] Describe a detailed introduction to the impact of ClimateTech on the economy

[135] Describe an introduction as we move to the last technology that will shape our future: 3D printing. How will it shape our future as 3D printing in itself is a convergence of different technologies

[136] Describe how 3D printing will disrupt globalization

[137] How will global trade based on services instead of production, made possible by 3D printing change societies?

[138] Describe how 3D printing will change global trade as it enables a global services industry instead of a production industry

[139] What questions should we ask to understand how 3D printing will shape our future?

How will 3D printing enable us to create custom and personalized products?

- How will 3D printing enable us to reduce waste and improve the sustainability of our production and consumption patterns?
- How will 3D printing enable us to develop new and innovative solutions to challenges such as climate change?
- How will 3D printing enable us to move from global to local production and consumption, and how will this change global trade?
- How will 3D printing affect the global economy, and what are the potential benefits and drawbacks of this technology?
- How will 3D printing impact the way we live, work, and interact with each other, and what are the potential social and ethical implications of this technology?

[140] Describe an introduction to the benefits of 3D printing.

[141] How will 3D printing enable us to create custom and personalized

products?

[142] How will 3D printing enable us to reduce waste and improve the sustainability of our production and consumption patterns?

[143] Summarize the benefits of 3D printing as discussed

[144] Expand on customization and personalization

[145] Expand on Personalized and custom-fit clothing using 3D printing in the far future

[146] Expand the section on Customized and personalized living spaces

[147] How will 3D printing change our healthcare? --> Could 3D printing enable 3D printing of organs? → Expand section on 3D organ and tissue printing

[148] How will 3D printing change space exploration?

[149] How will 3D printing shape our future if it converges with the other technologies discussed in this book?

[150] Provide a summary of how 3D printing will shape our future, be excited about the possibilities

[151] Write an epilogue that serves as a comment on or a conclusion to all that has been discussed in this book --> expand and be more optimistic and excited about the future --> Describe how you experienced writing this short history of our future.

ABOUT THE AUTHOR

Dr Mark Van Rijmenam

Dr. Mark van Rijmenam is The Digital Speaker. He is a leading strategic futurist who thinks about how emerging technologies change organizations, society and the metaverse. Dr. Van Rijmenam is one of the leading voices in technology and known for his candid, educated and balanced views on how technology can benefit but also threaten society. He is the founder of the Digital Futures Institute, with a mission to ensure a thriving digital future for business and society. Van Rijmenam is an international keynote speaker and 5x author, including Think Bigger (2014), Blockchain (2018), The Organisation of Tomorrow (2019), and Step into the Metaverse (2022).

BOOKS BY THIS AUTHOR

Step Into The Metaverse: How The Immersive Internet Will Unlock A Trillion-Dollar Social Economy

An incisive glimpse into the future of the internet

In Step into the Metaverse: How the Immersive Internet Will Unlock a Trillion-Dollar Social Economy, future tech strategist, entrepreneur, and thought leader Dr. Mark van Rijmenam delivers a startlingly insightful discussion about how the world as we know it will fundamentally change as the physical and the digital worlds merge into the metaverse, impacting the everyday experiences of people, companies, and societies. The author maps out the extraordinary opportunities and challenges facing business leaders, consumers, regulators, policymakers, and other metaverse stakeholders trying to navigate the future of the Internet.

In the metaverse, you can be who you want to be, where you want to be, and companies and consumers are only restricted by their own creativity how they can benefit from the immersive internet. With engaging commentary on issues ranging from avatars, identity and digital fashion to non-fungible tokens (NFTs), blockchain and the economics of the metaverse, this book also offers:

- Discussions on the importance of an open and interoperable metaverse build on the web 3.0 paradigm if we want to reign in the

control of Big Tech over our identity, data and lives.
- Explorations of the enormous—and largely untapped—potential for metaverse entertainment, including gaming, music, media, and sports and how brands can engage with their customers in novel ways and how digital twins will change how we work and innovate.
- Considerations related to the dangers of an always-on, immersive internet, including data breaches, avatar imposters, mental health issues, corporate and state surveillance, and the need for metaverse law.

A fascinating read you won't be able to put down, Step into the Metaverse belongs in the hands of executives, managers, and other business leaders who play a role in digital transformation or execution. It's also an invaluable resource for anyone interested in the future of technology, the internet, and social interaction.

The Organisation Of Tomorrow: How Ai, Blockchain And Analytics Turn Your Business Into A Data Organisation

The Organisation of Tomorrow presents a new model of doing business and explains how big data analytics, blockchain and artificial intelligence force us to rethink existing business models and develop organisations that will be ready for human-machine interactions. It also asks us to consider the impacts of these emerging information technologies on people and society.

Big data analytics empowers consumers and employees. This can result in an open strategy and a better understanding of the changing environment. Blockchain enables peer-to-peer collaboration and trustless interactions governed by cryptography and smart contracts. Meanwhile, artificial intelligence allows for new and different levels of intensity and involvement among human and artificial actors. With that, new modes of organising are emerging: where technology facilitates collaboration between stakeholders; and where human-to-human interactions are increasingly replaced with human-to-machine and even machine-to-machine interactions. This book offers dozens of examples of industry leaders such as Walmart, Telstra, Alibaba, Microsoft and T-Mobile, before presenting the D2 + A2 model – a new model to help organisations datafy their business, distribute their data, analyse it for insights and automate processes and customer touchpoints to be ready for the data-driven and exponentially-changing society that is upon us

This book offers governments, professional services, manufacturing, finance, retail and other industries a clear approach for how to develop products and services that are ready for the twenty-first century. It is a must-read for every organisation that wants to remain competitive in our fast-changing world.

Blockchain: Transforming Your Business And Our World

The internet was envisaged as a decentralised global network, but in the past 25 years it has come to be controlled by a few, very powerful, centralised companies. Blockchain is a technological paradigm shift that allows secure, reliable, and direct information transfer between individuals, organisations, and things, so that we can manage, verify, and control the use of our own data.

Blockchain also offers a new opportunity for humanity to fix some major problems. It can authenticate data, manage its analysis, and automate its use. With better data comes better decision-making. In this way, Blockchain can contribute to solving climate change, reduce voting fraud, fix our identity systems, improve fair trade, and give the poor an opportunity to improve their lives by monetising their (digital) capital. A world built upon peer-to-peer transactions and smart contracts can empower individuals and communities.

This book offers a fresh perspective with which to consider this transformative technology. It describes how Blockchaincan optimise the processes that run our society. It provides practical solutions to global problems and offers a roadmap to incorporate Blockchain in your business. It offers a blueprint for a better world. Filled with easy-to-understand examples, this book shows how Blockchain can take over where the internet has fallen short.

Think Bigger: Developing A Successful Big Data Strategy For Your Business

Big data--the enormous amount of data that is created as virtually every movement, transaction, and choice we make becomes digitized--is revolutionizing business. Offering real-world insight and explanations, this book provides a roadmap for organizations looking to develop a profitable big data strategy...and reveals why it's not something they can leave to the I.T. department.

Sharing best practices from companies that have implemented a big data strategy including Walmart, InterContinental Hotel Group, Walt Disney, and Shell, Think Bigger covers the most important big data trends affecting organizations, as well as key technologies like Hadoop and MapReduce, and several crucial types of analyses. In addition, the book offers guidance on how to ensure security, and respect the privacy rights of consumers. It also examines in detail how big data is impacting specific industries--and where opportunities can be found.

Big data is changing the way businesses--and even governments--are operated and managed. Think Bigger is an essential resource for anyone who wants to ensure that their company isn't left in the dust.

Made in United States
Troutdale, OR
11/03/2026

56844904R00080